MW01221706

Dedicated to
Bosch van
Oudtshoorn,
pioneer researcher
of the chemistry
and pharmaceutical properties of
South African aloes.

Supported by the
National Botanical
Institute

BRIZA
PUBLICATIONS

Published by
Briza Publications
CK90/11690/23
P.O. Box 56569, Arcadia, 0007
Pretoria
South Africa

First edition 1996

English edited by Aïda Thorne
Cover design by Andrew Breebaart
Typesetting by Briza Publications
Reproduction by Unifoto, Cape Town
Printed and bound by Tien Wah Press, Singapore

Cover photo: *Aloe dichotoma* by Pitta Joffe – NBI
Cover background photo: *Aloe aculeata* by Eben van Wyk
Back cover photo: *Aloe comptonii* by Marinda Koekemoer – NBI
Spine photo: *Aloe barberae* by Pitta Joffe – NBI
Photo on this page: *Aloe sessiliflora* by Duncan Butchart

ISBN 1 875093 04 4

Contents

Acknowledgements

Without the generous contributions by many aloe enthusiasts, specifically towards providing photographic material, the production of this book would have been impossible.

The publisher and authors wish to thank the following institutions and persons:

- The National Botanical Institute of South Africa for permission to make use of their photographic material in the guide

- Members of the Succulent Society of South Africa and others for their quick response and thousands of colour slides they submitted for publication

- The following individuals need a special word of thanks for their contributions towards photographs, locating rare aloes and general liaison (alphabetic order): Charles Craib, Bernhard de Souza, Andy de Wet, Dave Hardy, Pitta Joffe, Brian Kemble, Sanet Krynauw, Cameron McMaster, Wally Menne, Geoff Nichols, Ronel Niemand, Jo Onderstall, Leo Thamm, Frans van der Merwe, Ernst van Jaarsveld, Eben van Wyk and Fanie Venter.

From a production point of view we would like to thank Ilse van Oudtshoorn for the illustrations and Aïda Thorne for English editing.

Frits van Oudtshoorn
Ben-Erik van Wyk
Gideon Smith

Introduction

Aloes form a conspicuous part of the South African landscape and their striking beauty is appreciated by many people. This book is a **user-friendly field guide to all South African aloes**. Short descriptions and clear photographs of all 125 species occurring within the boundaries of South Africa have been included, showing the plants in habitat and also most of their distinguishing characteristics. In this regard it is aimed at as wide an audience as possible. Botanists, field naturalists, students of our enormous succulent flora and gardeners with little more than a passing interest should now find aloe identification a far simpler task.

Distribution maps of all the species are provided for the first time. This information is often very helpful as an aid to identification, but descriptions and photographs should also be consulted. Only the South African (including Swaziland and Lesotho) distribution ranges are given for those species that occur further afield in neighbouring countries.

To keep disruptive name changes to a minimum, we have mostly followed the classical species concepts of the well-known aloe authority, Dr G.W. Reynolds, first published in 1950. Due to new developments in the classification of aloes, some changes were unavoidable. It is important to note that there are no absolute criteria for deciding on what rank to apply to a particular aloe and how different it should be from other aloes to justify the rank of species. There are many examples of regional forms that have been described as species. If these are reasonably recognisable as distinct entities, we have retained the existing names.

For practical reasons, the ten groups into which the aloes are placed in this book are based mainly on size, growth habit and branching pattern. Closely related species which differ in their growth habit may therefore have ended up in different parts of the book. A more natural grouping would have been possible (and would indeed have been more satisfying to botanists), but our main intention was to help even the most inexperienced aloe lover find the correct name. The groups are given more or less in a descending sequence based on the size of mature plants, that is from the tree aloes (Group 1) to the grass aloes (Group 10). Once experience is gained in recognising the groups, selecting the correct species from within the group will be comparatively easy.

The discussions of the species and distribution maps can hardly be regarded as perfect because there is still so much to learn about aloes. However, we do believe that this book fills a void that has existed for a number of years and that aloe identification will now be considerably easier.

ALOE-LIKE PLANTS

Probably the two most conspicuous characteristics of aloes are their succulent leaves that are arranged in rosettes, and their tall, usually candle-like inflorescences. Aloes are therefore generally easy to distinguish from their relatives, even though these two and their other characteristics vary widely. For example, aloes can attain tree-like dimensions with massive fibrous trunks of up to 20 m tall or they could be miniatures of no more than a few centimetres high. The flowers are always more or less tubular and are borne on a simple or branched inflorescence. Additionally, the leaves are usually armed with fierce or soft marginal and terminal prickles and remain on the plant when they dry out.

Aloes have traditionally been grouped with the lily-like plants (the family Liliaceae) along with the true lilies, tulips and many other genera and species. Following recent refinements to the classification system, the aloes and their close relatives, such as the red-hot pokers and bulbines, are placed in a family of their own, the Asphodelaceae. Another point of view is that all the plants with succulent leaves, such as the aloes, gasterias and haworthias, should be placed in their own family, the Aloaceae.

Although fairly easy to recognise, the aloes are sometimes confused with a number of other plants that also occur in South Africa. These are briefly described below, with emphasis on differences amongst them.

Agave
The only exotic genus widely cultivated in South Africa (and in fact all over the world) with which aloes could be confused is the agaves (Century plants or Garingbome). Two species of this New World genus, *Agave americana* and *A. sisalana*, are naturalised in the subcontinent. These species can easily be separated from aloes by virtue of their monocarpic habit, that is they die after having flowered. This usually takes quite a few years or even decades, and is the characteristic from which the common name, Century plant, was derived. The leaves of the agaves are firmer and tougher than those of aloes and more inclined to be sword-shaped. This means that they usually become thinner towards their bases, with an abrupt thickening at the point of attachment to the stems. In addition, the inflorescences are usually several metres high and the dull greenish-yellow flowers of most species have a completely different shape.

Kniphofia
In contrast to the aloes, the *Kniphofia* species (Red-hot pokers or Vuurpyle) have non-succulent leaves that in most cases have a distinct ridge or keel along the lower surface. The inflorescences are invariably unbranched and the narrow, tubular flowers are generally tightly packed within the relatively short racemes. Kniphofias usually grow in wetter habitats than aloes.

Bulbine
The species of *Bulbine* always have open, non-tubular, yellow, orange or rarely white flowers, which are remarkably uniform in their star-shaped appearance. A

Gideon Smith

Agave americana

Adela Romanowski – NBI

Agave angustifolia

Frits van Oudtshoorn

Kniphofia hybrid

Frits van Oudtshoorn

Kniphofia hybrid

Ben-Erik van Wyk

Bulbine natalensis

Pitta Joffe – NBI

Bulbine frutescens

Gideon Smith

Gasteria acinacifolia

Ben-Erik van Wyk

Gasteria bicolor

unique characteristic of *Bulbine* is the distinctly hairy filaments. (The filaments are the stalks carrying the pollen-bearing anthers.) Most species of *Bulbine* are rather small and insignificant and cannot be used effectively in general horticulture. However, one species, *B. frutescens* (Cat's tail or Balsemkopiva), is used extensively as a border plant or ground cover.

Gasteria

More closely related to the aloes than the above genera are the sixteen species of *Gasteria* which are characterised by their pointed, yet spineless leaves and their distinctly curved, basally bulbous flowers. The leaves of most gasterias are distinctly mottled with white and they often have a ridge or keel on their lower surfaces. Gasterias generally grow in the shade of surrounding vegetation and they rarely have stems as do aloes.

Haworthia

Species of the genus *Haworthia* have the general appearance of miniature aloes. The most important distinguishing characteristics of this distinctive group are the dull, whitish, two-lipped flowers and the small stature of the plants. Like the gasterias, these species generally prefer shady growing positions.

Astroloba

The few species of *Astroloba* are closely related to and should arguably be included in *Haworthia*. Astrolobas form short, thin, often creeping stems that are densely packed with small, triangular leaves. In contrast to *Haworthia*, the flowers are regular, that is not two-lipped, and sometimes carry longitudinal ribs of inflated tissue.

Poellnitzia

Poellnitzia has only a single species, the stems of which are virtually identical to those of *Astroloba*. However, the flowers are bright red and are carried erectly on a horizontal flowering stem. Furthermore, the flowers of *Poellnitzia* are unlike those of any other aloe-like plant in that they do not open in the true sense of the word.

Chortolirion

There is only one species in this genus. The flowers are very similar to those of *Haworthia*, but the leaves die back in winter and the plant has a distinct underground bulb. It occurs primarily in the grasslands of the Southern African interior.

Pitta Joffe – NBI

Haworthia attenuata

Ben-Erik van Wyk

Haworthia arachnoidea

Ernst van Jaarsveld – NBI

Astroloba sp. nov.

Ernst van Jaarsveld – NBI

Astroloba foliolosa

Ben-Erik van Wyk

Poellnitzia rubriflora

Ben-Erik van Wyk

Poellnitzia rubriflora

Gideon Smith

Chortolirion angolense

Gideon Smith

Chortolirion angolense

Medicinal uses

The juice of aloes has been used medicinally for centuries. It is said that Alexander the Great conquered the island of Socotra to gain control over the main supply of aloetic medicine. To this day, most of the drug aloes of the world are produced from the famous *Aloe vera*, sometimes still known by the incorrect name *Aloe barbadensis*. This species originally came from North Africa or Arabia, but is widely cultivated in many parts of the world, particularly in the West Indies. Commercially, the product is known as Curacao Aloes or Barbados Aloes.

In South Africa, *Aloe ferox* is the source of a similar medicinal product known as Cape Aloes. This purgative drug is used for stomach complaints, mainly as a laxative to 'purify' the stomach. It is a dark brown, resinous solid and has been commercially utilised for centuries. Most of the annual production is exported to Europe, but a substantial quantity is also marketed and used locally. It is a major ingredient of various traditional medicines and is also sold in its original form or as a powder.

When an aloe leaf is cut, an extremely bitter yellow juice oozes from small canals situated just below the surface in the green part of the leaf. The fresh juice is widely used as a first-aid treatment for burn wounds, while the inner leaf pulp is processed into aloe gel (see page 14). To produce the medicinal product called Cape Aloes, the yellow juice is collected and dried by an age-old method. An animal skin or plastic sheet is placed in a hollow in the ground to collect the leaf exudate. Leaves are harvested with sickles and then carefully stacked around the hollow, with the cut surfaces facing inwards. The leaves in such a typical stack are kept together by the thorns along the edges which prevent them from slipping. Experienced harvesters never remove more than about eight leaves at a time, so that the plant will recover.

The exudate is boiled in drums on open fires to remove the water, after which it solidifies to form a dark brown solid, the so-called aloe lump. The active ingredient is a chemical compound known as aloin (sometimes also called barbaloin). For export quality Cape Aloes, the product should contain at least 18 percent aloin.

Duncan Butchart

Aloe ferox, the source of Cape Aloes

Peter Winter

Harvested *Aloe ferox*

Peter Winter

Stack of leaves

Ben-Erik van Wyk

Dried leaf sap (Aloe lump)

Ben-Erik van Wyk

Aloe ferox medicinal products

Cosmetic uses

In recent years, there has been a growing interest in aloe products, particularly aloe gel for the cosmetics industry. The source of most of the products is the well-known exotic species, known botanically and commercially as *Aloe vera*. It is a stemless aloe which produces side-shoots and gradually forms dense clusters. The leaves are thick and fleshy, with small, harmless teeth along the margins. The flower colour varies from red to yellow. This aloe originally came from North Africa or Arabia, but spread into the Mediterranean area many centuries ago. From here the plant was introduced into Central America (probably by the Spaniards), and it is now running wild in many areas, from the West Indies to Bolivia. A lucrative industry has developed in the West Indies, where substantial quantities of aloe gel are produced from cultivated plants. The plants are easily propagated from suckers and provide an important source of income for rural communities. Other production areas include East Africa, north-western India and southern China.

To produce the valuable product known as aloe gel or aloe pulp, the fleshy inner portion of the leaf (the so-called white juice) is carefully separated from the outer green part which contains the bitter yellow exudate. The fleshy part is tasteless, and is simply crushed and homogenised to form the gel, or it may be extracted chemically and spray-dried to form gel powder.

Aloe gel is a watery mixture of pectic carbohydrates (jelly-like polysaccharides), amino acids, minerals, trace elements, organic acids and various minor compounds such as enzymes, claimed to be 'biologically active'. The gel is used in the health food industry to produce tonics, but the main commercial application is in the lucrative cosmetics industry. The popularity stems mainly from the worldwide move towards more natural products that are free from artificial chemicals. A wide range of hair and skin care products is manufactured, including ointments, soaps, shampoos, skin moisturising creams and sun tan lotions. It is now also possible to buy cosmetics made from the leaf gel of our indigenous *Aloe ferox*.

Aloe vera

Alice Biemond

Aloe vera cosmetic products

Ben-Erik van Wyk

Aloe ferox cosmetic products

Ben-Erik van Wyk

Conservation

With few exceptions, aloes are protected by environmental legislation in all the provinces of South Africa. It is therefore illegal to remove plants from their natural habitats, unless one is in possession of a collecting permit issued by a provincial nature conservation authority, and has the consent of the land owner. Although some species have never had large populations, various factors have contributed to the decline in numbers of several species. These factors include urban and industrial expansion, agricultural development, afforestation and mining activities. Unfortunately, a number of aloes are also threatened by extensive collection and under no circumstances should plants be taken from the wild illegally. Nurseries specialising in indigenous and/or succulent plants should be sourced for propagation material.

The conservation status of all the aloes of South Africa is mentioned under each species. The definitions for the various categories given below are widely accepted by conservation authorities.

Endangered: Plants that are in immediate danger of becoming extinct if the factors causing their decline continue to operate. Included here are plants whose numbers have been reduced to a critical level or whose habitats have been so drastically reduced that they are deemed to be in immediate danger of extinction.

Vulnerable: Plants believed likely to move into the Endangered category in the near future if the factors causing their decline continue operating.

Critically rare: Plants with small populations that are not at present Endangered or Vulnerable, but are at risk as some unexpected threat could easily cause a critical decline. These plants are usually localised within restricted areas or are thinly scattered over a more extensive range. The category is termed Critically rare to distinguish it from the more generally used word 'rare'.

Indeterminate: Plants known to be Extinct, Endangered, Vulnerable, or Critically rare but where there is not enough information to say which one of the four categories is appropriate.

Insufficiently known: Plants that are suspected but not definitely known to belong to any of the above categories, because of the lack of information.

Not threatened: Used for plants which are abundant or no longer in one of the above categories due to increases in population sizes or to subsequent discoveries of more individuals or populations.

Ben-Erik van Wyk

Endangered – *Aloe pillansii*

Graham Williamson

Vulnerable – *Aloe pearsonii*

Ben-Erik van Wyk

Critically rare – *Aloe peglerae*

NBI

Endangered – *Aloe polyphylla*

Kobus Venter

Vulnerable – *Aloe krapohliana*

Ben-Erik van Wyk

Critically rare – *Aloe comosa*

Cultivation

Aloes must rate as some of the most rewarding South African plants to cultivate. The stark beauty of their often strange and inspiring architectures makes them suitable as accent plants in a variety of settings. However, they need not only be used as feature plants, for example in containers around a swimming-pool or on a patio, but should increasingly find their way into general gardening. They are pre-adapted to the often harsh South African climate and are compatible with a wide variety of horticulturally popular exotic and indigenous plants. To effectively show off the interesting shapes and sizes of aloes, it is not necessary to establish a conventional rockery. They can be used in almost any setting and in conjunction with most common garden plants. With the large variety of different growth forms found amongst aloes – from large trees to shrubs and ground covers – it is therefore possible to select suitable species for any horticultural purpose.

One hardly requires green fingers to be successful in cultivating aloes. Some of the common species can be readily grown almost anywhere in the country, from the coast to the arid South African interior and even above the often climatically severe Great Escarpment. To grow certain species successfully will require protection from out-of-season rainfall or too high light intensity. And then there are of course some species that will always be difficult or even downright impossible to grow successfully. These are in the minority and the average gardener should not even attempt to grow them. Here *Aloe haemanthifolia* immediately comes to mind.

In general, aloes grow well in warm climates in soils that have adequate drainage. Only a few species can withstand severe frost and protection may be necessary in cold climates. Most tree and shrub-like species are cultivated from stem cuttings which are allowed to dry for a few days, and are then planted directly at the spot intended. More or less the same applies to the suckering species that form large clumps of numerous rosettes. In the latter case, small plantlets can be removed from the mother plant with a sharp knife and be left to dry, or if roots are already present, the rosettes can be planted in their final positions. In both cases plants grown in this way will reach maturity more rapidly than plants grown from seed. Propagation from seed is relatively easy and should be used more often, particularly to ensure the survival of some of the rare species. In addition, it is very rewarding to grow aloes in this way.

Aloes, once established, require very little after-care. If planted in rich soil that is regularly mulched with well-decomposed compost, these plants can be left undisturbed for many years. It is a fallacy that watering should be withheld for aloes to grow optimally. Aloes can indeed tolerate long periods of drought but they will thrive and flower better if adequate water is provided in the correct season. Healthy, well-grown plants are also more resistant to insect attacks.

Aloe thraskii x *A. ferox*

Andy de Wet / Leo Thamm

Aloe petricola x *A. arborescens* x *A. marlothii*

Andy de Wet / Leo Thamm

Variegated form of *Aloe arborescens*

Ben-Erik van Wyk

Pests and diseases

Prevention is better than cure. A thriving, vigorous plant will be less prone to pests and diseases than one which suffers from incorrect watering, poor drainage or too much shade.

However, like most plants, aloes are also prone to a variety of diseases. Few of these will lead to the rapid demise of the plants, but will certainly spoil their appearance. Only some of the more obvious and common pests are briefly mentioned here.

One of the most unsightly aloe infestations is the one caused by white scale insects that gather in neat white rows on the leaves, especially the lower surfaces. These and other scale insects can easily be killed with modern aerosol contact insecticides. If left untreated, the insects will eventually cover the entire plant and it may die.

Another unsightly problem is the occurrence of aloe cancer (also called galls) that causes severe deformation of the leaves or inflorescences. This problem results from mite infestations that stimulate unnatural cell growth. It is not easy to control this disease and it is often better to destroy the infected plants, taking care not to infect other plants. As a desperate measure in the case of a rare or valuable plant, the infected areas should be removed with a sharp knife and the wounds treated with insecticide, applied directly in the form of a wettable powder. Some species, such as *Aloe longistyla* and some forms of *A. arborescens* are particularly prone to this disease.

Round brown spots are sometimes seen on aloe leaves. This is caused by a fungus and the disease is generally referred to as aloe rust. These spots spoil the appearance of the plant and unfortunately recovery is very slow, even when treated with fungicide.

Probably the worst pest in South Africa is the aloe snout beetle. This insect tunnels into the heart of the crown where it lays its eggs. The larvae hollow out the stems, so that the plants start to rot and ultimately collapse. Damage is unfortunately often detected only when it is too late. Plants can be treated by injecting neat systemic insecticide into small holes drilled into the stem.

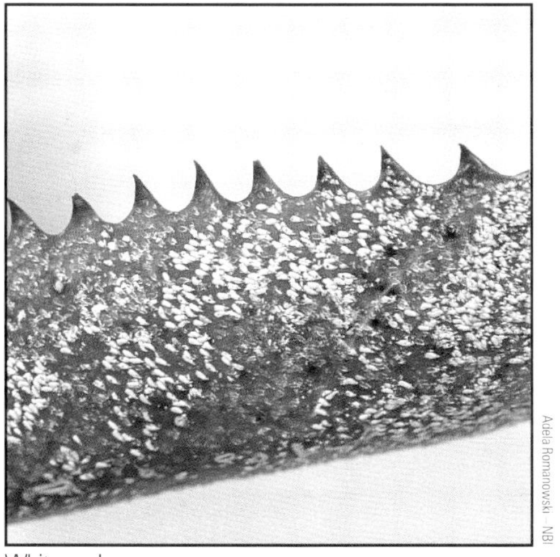
Adela Romanowski – NBI

White scale

Adela Romanowski – NBI

Aloe cancer

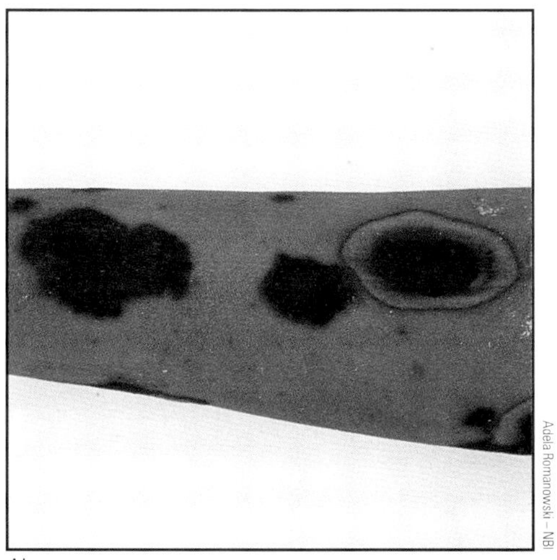
Adela Romanowski – NBI

Aloe rust

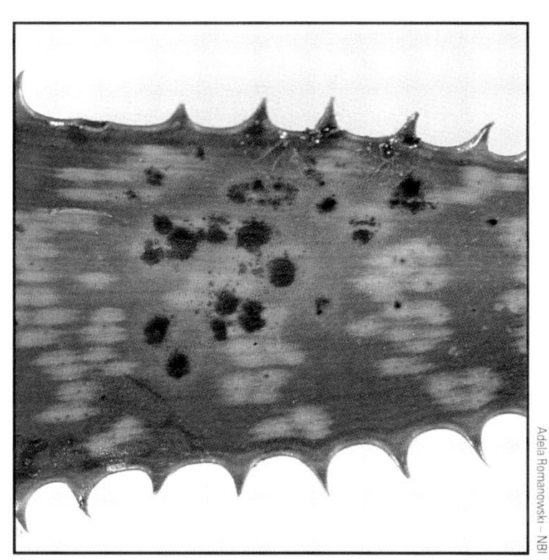
Adela Romanowski – NBI

Aloe rust

Ernst van Jaarsveld – NBI

Snout beetle

Ernst van Jaarsveld – NBI

Snout beetle larva

Hybrids

Aloe hybrids are commonly encountered and may cause problems with identification if one is unaware of the fact that most aloes will interbreed freely. There are so many hybrids that it was impossible to include them all in this book. In the natural habitat it is generally easy to recognise hybrids because they are usually intermediate between the two parents, both of which should be present nearby. In contrast, it may be impossible to identify garden hybrids since it may not be obvious which plants were the parents. One of the best-known natural hybrids is a cross between *Aloe ferox* and *A. arborescens*. Large numbers are found at various localities where the distribution areas overlap, and this striking cross is also commonly seen in gardens.

Apart from the large variety of pure species to choose from for cultivation, numerous *Aloe* hybrids are now becoming popular in general gardening. Their popularity is due to the fact that they are often horticulturally more beautiful than either of the parents. Hybrids often grow more rapidly, flower sooner and produce more striking flowers than the pure species. This phenomenon is known as 'hybrid vigour' and makes some of the hybrids highly sought-after for cultivation.

The immense potential of hybrid aloes as garden plants was recognised many decades ago through the foresight of Mr At Koeleman. He was eventually instrumental in establishing the South African *Aloe* Breeders' Association, which is the international body for the registration of *Aloe* cultivars. This group of dedicated aloe breeders is currently involved in selecting and improving artificially produced *Aloe* hybrids for use as cultivars. The plants are generally selected for their longer flowering periods and more striking flowers and foliage.

In a carefully controlled hybridisation programme, parent plants which have the desired characteristics are selected for the first-generation crosses (e.g. small stature, large flowers). Through carefully studying the inheritance of characteristics in the offspring and then selecting those plants in which the most desirable features are best established, it is possible to obtain hybrids that fulfil the needs of the end-grower.

Some of these natural hybrids and cultivars are illustrated on the facing page.

Ben-Erik van Wyk

Natural hybrid — *Aloe arborescens* x *A. ferox*

Andy de Wet / Leo Thamm

Cultivated hybrid — *Aloe arborescens* x *A. marlothii* x *A. petricola*

Andy de Wet / Leo Thamm

Cultivated hybrid — *Aloe arborescens* x *A. suprafoliata* cultivar Vinkel

Layout and terminology used in this book

To facilitate identification, all 125 South African aloes are arranged in ten groups, based mainly on growth habit. The sequence starts with the tree aloes and ends with the grass aloes, that is from large to small. The groups are often artificial and species are grouped together merely for convenience. In some groups, however, the species are truly related, such as the rambling aloes (Group 4), creeping aloes (Group 5), speckled aloes (Group 7) and grass aloes (Group 10). It should be borne in mind that the boundaries between some groups are not sharp and in certain cases a species may well fit comfortably into more than one group. In such cases the plant has been included in the group with which it can be more easily associated. For example, *Aloe suffulta* is treated as a spotted aloe because of the spotted leaves, even though this arrangement disrupts an otherwise natural group.

Each of the species are treated in the following sequence: (1) **brief description**; (2) **flowering time**; (3) **distinguishing characteristics**; (4) **geographical distribution**; (5) **conservation status**; and (6) **general notes**. As far as possible the descriptions of the species have been written in non-technical terms so that anyone would be able to understand them. However, a brief pictorial glossary is included here for terms that warrant further explanation. Every species is usually represented by at least two photographs showing distinctive features, such as habitat, growth form, leaves and flowers.

Distribution maps of all the species have been included as an aid to identification. When plants are studied in their natural habitat, the number of options can immediately be reduced and finding the correct name for a species should be relatively easy.

Illustrated glossary

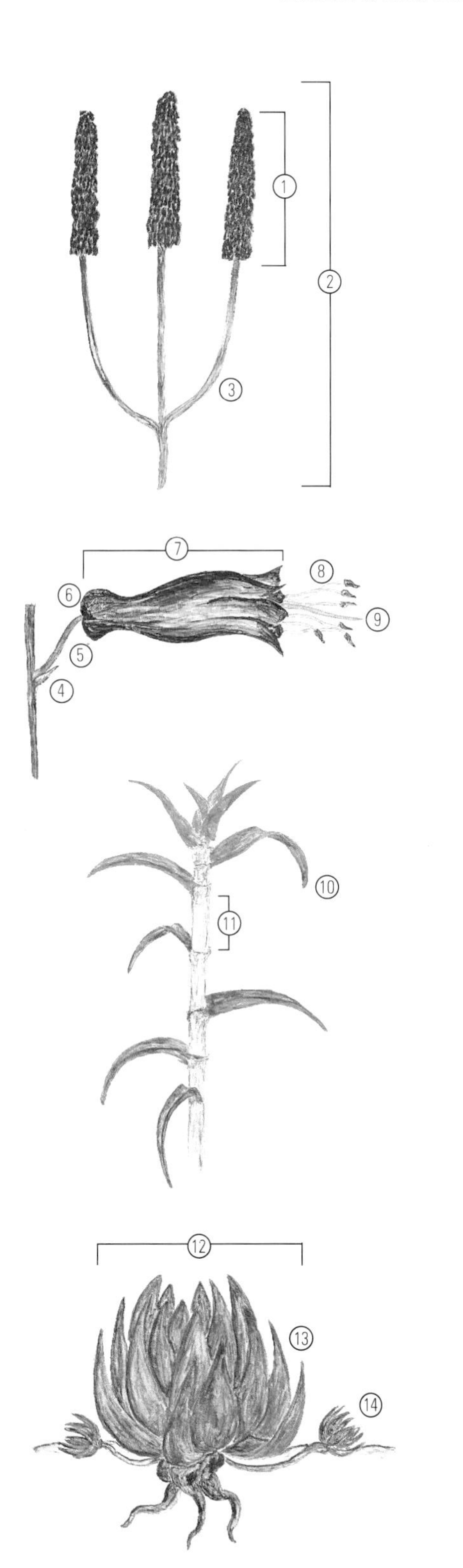

① raceme

② inflorescence

③ peduncle

④ bract

⑤ flower stalk

⑥ basal swelling

⑦ perianth

⑧ exserted stamens

⑨ style and stigma

⑩ recurved leaf

⑪ internode

⑫ rosette

⑬ incurved leaves

⑭ sucker

THE GROUPS

Group 1 TREE ALOES
5 species
Page 28
Tree-like appearance
Main trunk with side-branches
Small rosettes
Dead leaves not persistent

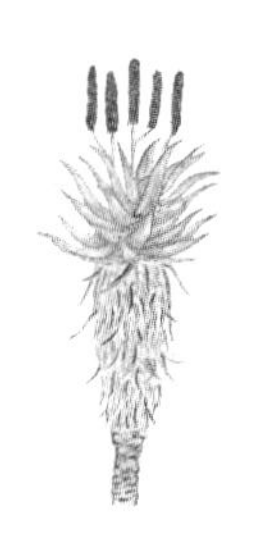

Group 2 SINGLE-STEMMED ALOES
15 species
Page 40
Single, erect main stem
Sometimes branched in the upper part
Large, single rosette or several if branched

Group 3 MULTISTEMMED ALOES
9 species
Page 72
Medium to large shrubs
Multistemmed at or near ground level
Leaves usually strongly recurved or curved upwards

Group 4 RAMBLING ALOES
5 species
Page 92
Thin and slender, almost wiry stems
Slightly fleshy leaves
Widely spaced leaves
Sheathing portion of leaves usually striate
Racemes short with few flowers

Group 5 CREEPING ALOES
7 species
Page 104
Creeping stems
Rosettes tilted to one side
Dull greenish leaf colour
Harmless white teeth
Head-shaped, many-flowered racemes
Flowering in summer

Group 6 STEMLESS ALOES
23 species
Page 120
Stem short or absent
Rosettes borne close to the ground
Rosettes single or rarely in small groups

Group 7 SPECKLED ALOES
5 species
Page 168
Stem usually short
Numerous small, round spots on both leaf surfaces
Southern and western parts of the Cape

Group 8 SPOTTED ALOES
26 species
Page 180
Stem short or absent
Leaves copiously spotted
Spots usually oblong or H-shaped
Flowers conspicuously inflated at the base

Group 9 DWARF ALOES
6 species
Page 234
Rosettes small
Rosettes rarely single
Leaves usually narrow and incurved
Leaves often with raised white tubercles

Group 10 GRASS ALOES
24 species
Page 248
Plants slender and more or less stemless
Leaves long and narrow
Leaves only slightly succulent
Lower leaf surfaces often with whitish spots
Inflorescences invariably single

*T*REE ALOES

Group 1

Easily recognised by their **tree-like appearance**, these species form one of the most distinctive and interesting groups of aloes. The tree growth form is unusual amongst the aloes, since most of them lack a **main trunk with side-branches**. In addition, the leaves are borne in **small rosettes** and **dead leaves are not persistent** on the stems as in most other species. Geographically the tree aloes occur in three separate and very different areas: the arid north-western corner of the Cape (*A. dichotoma*, *A. pillansii* and *A. ramosissima*), the south-western Cape (*A. plicatilis*), and the subtropical eastern and south-eastern parts of South Africa (*A. barberae*).

SPECIES

A. barberae
A. dichotoma
A. pillansii
A. plicatilis
A. ramosissima

Marike Bruwer

Aloe ramosissima

Aloe barberae

Plants occur as solitary, much-branched trees of up to 18 m high and could form trunks of up to 3 m in diameter. Branching is either two or three-forked. The bark of the tree trunks is greyish-brown and rough to the touch. The leaves are deeply channelled and recurved. Leaf margins are armed with small, firm, whitish teeth. The branched inflorescence is about 500 mm long and usually consists of up to three racemes which branch off from the peduncle fairly low down. The inflorescences rarely exceed the rosettes in length and are often hidden amongst the leaves. The flowers are wider in the middle than at the ends, making them appear rather swollen. Flower colour ranges from salmon-pink to orange.

The flowering period is from June to August.

The height of mature specimens and the flower colour immediately separate *A. barberae* from the other tree aloes – it is the tallest of them all and the only one with orangy-pink flowers. Young trees of *A. barberae* can easily be distinguished from *A. dichotoma* by their much larger, bright green and deeply channelled leaves.

Aloe barberae is essentially a forest dweller and favours dense, tall bush and low forests. It occurs in a broad coastal zone that stretches slightly inland from East London in the Eastern Cape Province northwards through KwaZulu-Natal, Swaziland and Mpumalanga to southern and central Mozambique.

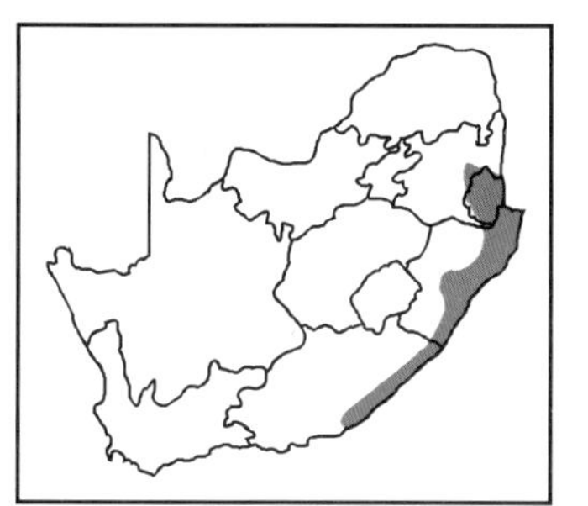

The species is not threatened.

Unfortunately the two names *A. bainesii* and *A. barberae* were published simultaneously for the same plant. In a long-overlooked scientific paper that appeared shortly after these names were established, *A. barberae* was explicitly chosen as the name to be used and was recently reinstated for this reason. The scientific name refers to Mary Elizabeth Barber (née Bowker), one of the pioneer plant collectors of South Africa. The national tree number is 28. Common names are Tree aloe, Boomaalwyn and Mikaalwyn. In Zulu, the species is known as inKalane enkulu, that is 'the big one'.

It is easy to cultivate and can be propagated by cuttings or truncheons. Plants will grow in almost any climate and soil type, but should be protected from severe frost and drought. It is noteworthy that the leaves of young specimens are longer and more widely spaced than those of mature specimens. A small form of *A. barberae*, about 2 m tall at maturity, has been recorded from Mozambique. This form is not common in cultivation.

GW Reynolds – NBI

Craig Hilton-Taylor

Pitta Joffe – NBI

Aloe dichotoma

This distinctive tree aloe has smooth branches, but the bark on the trunk forms large golden-brown scales with razor-sharp edges. The crown is often densely rounded as a result of the repeatedly forked branches. The somewhat blue-green leaves are borne in relatively small, terminal rosettes. Each leaf is about 50 mm wide at the base, 300 mm long, narrow and oblong in shape, with inconspicuous teeth along the margins. In juvenile plants the leaves are ranked in vertical rows, but in older plants they gradually become spirally arranged in rosettes. The inflorescence is short, erect and branched with three to five racemes. The bright yellow flowers are comparatively large, but short and swollen in the middle.

Aloe dichotoma flowers from June to August.

Some sparsely branched forms of *A. dichotoma* resemble *A. pillansii* in general appearance, but can easily be distinguished by the much smaller rosettes and erect racemes. Some trees with extensively branched, dense crowns are almost identical to *A. ramosissima*, but can be differentiated on account of the presence of at least a short trunk.

This species is a conspicuous component of the arid parts generally known as Namaqualand and Bushmanland. It occurs in rocky areas, from near Nieuwoudtville northwards into Namibia and eastwards to Upington and Kenhardt.

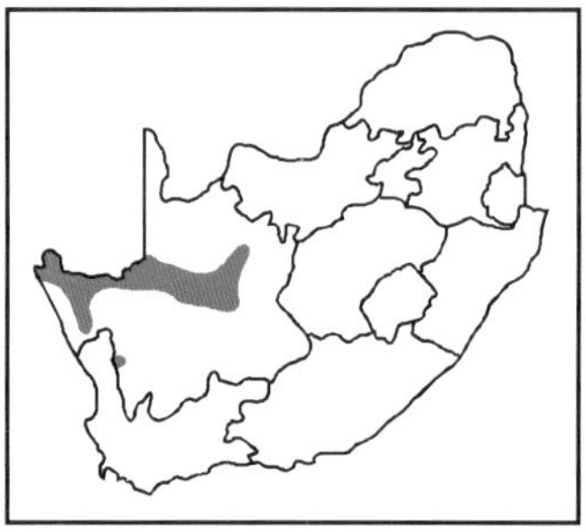

The species is not threatened.

The common name of *A. dichotoma*, Quiver tree or Kokerboom, is derived from the age-old use of the hollowed stems for quivers by the San people. The original name appears to be Choje. This species is a distinctive feature of the landscape in the north-western parts of the Cape, and the so-called quiver tree forests found at some localities will make an indelible impression on any visitor. The scientific name refers to the forked branches ('dichotomous' means 'forked'). The national tree number is 29. *Aloe dichotoma* is easily grown from seed and will grow fairly rapidly under suitable conditions.

Forsyth van Nierop

Duncan Butchart

Richard Warmough

Aloe pillansii

This stately aloe with robust, erect (never spreading) branches carries large rosettes of greyish-green leaves at the terminal ends of the stems. The leaves distinctly clasp the branches and are fairly large, being up to 600 mm in length and about 100 mm wide at the base. The leaf margins are armed with small white teeth. The inflorescence is branched, arising from the lowest leaves of the rosettes and each having numerous racemes. The flowers are yellow and slightly swollen in the middle.

Aloe pillansii flowers in October.

In habitat, *A. pillansii* cannot be confused with any other tree aloe; the plants are much more robust and have fewer, yet larger rosettes. The inflorescences that arise almost horizontally from the lower leaves distinguish this species from *A. dichotoma*, where the inflorescence is less extensively branched, erect and borne higher up. The bark is somewhat coarse in old specimens but lacks the scaly appearance of *A. dichotoma*.

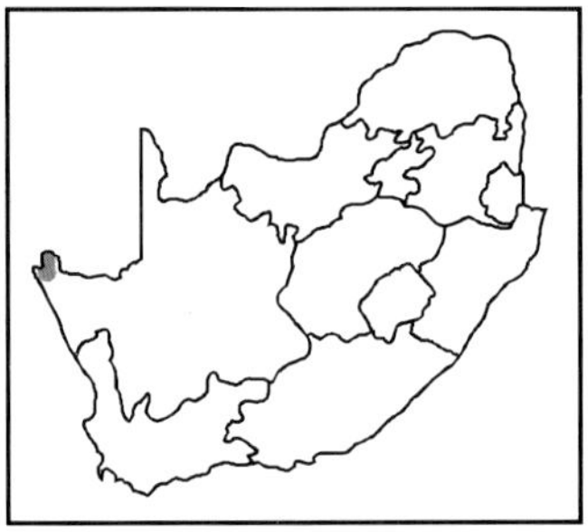

Aloe pillansii occurs from Cornell's Kop in the Richtersveld northwards to the Brandberg in Namibia.

Aloe pillansii is regarded as endangered due to overgrazing and collection.

This species leaves a lasting impression on anyone who has seen it in its natural environment. It qualifies as one of the most exceptional botanical features of South Africa, and at least in grotesque shape is comparable to the redwoods of the USA. The common name for this species is Giant quiver tree or Reusekokerboom. The scientific name commemorates Neville S. Pillans, a well-known Cape botanist, who first collected this remarkable species. The national tree number is 30.

George Lubbers

Geoff Tribe

Ben-Erik van Wyk

Ben-Erik van Wyk

Aloe plicatilis

This unique, striking tree aloe is one of the botanical treasures of South Africa. The stems are forked and up to 5 m high in large specimens, with clusters of strap-shaped leaves arranged in two opposing rows, each cluster resembling an open fan (hence the common name Fan aloe). Individually the leaves are grey-green in colour, and about 300 mm long and 40 mm wide. The racemes are cylindrical in shape and are always single in each leaf cluster (never branched). They have up to thirty tubular, scarlet flowers, each about 50 mm long and somewhat fleshy in texture.

This beautiful aloe flowers in spring, from August to October.

The unusual arrangement and shape of the leaves make this a unique species amongst the tree aloes. *Aloe plicatilis* has indeed no known relatives within the genus *Aloe*. The leaves produce copious amounts of a watery exudate, which soon crystallises to a pale yellow wax-like solid. The species is superficially similar to the equally unusual *A. haemanthifolia*, but the latter is a small, stemless plant.

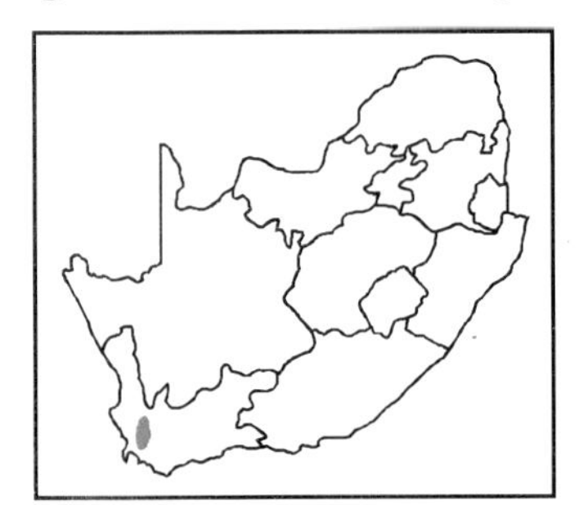

Aloe plicatilis is found only in the western Cape mountains of South Africa, from Franschhoek in the south to Elandskloof in the north. The plants grow on steep rocky slopes in well-drained, acid sandy soils. It is one of only a few aloes found in fynbos vegetation, where it grows in high rainfall areas with an annual precipitation of 600 to 1 200 mm and more.

Aloe plicatilis is not threatened, but has a limited natural distribution area and should be protected.

The Fan aloe or Waaieraalwyn is relatively easy to cultivate and makes a wonderful feature in any garden. Additional attention to watering is required in the summer rainfall area. The scientific name somewhat inaccurately refers to the fan-shaped rosettes. The national tree number is 29.6.

Ben-Erik van Wyk

Craig Hilton-Taylor

Ben-Erik van Wyk

Aloe ramosissima

This shrubby plant is probably the most profusely branched of all aloes. The stems are smooth and end in relatively small rosettes of narrow oblong leaves, each of which is up to 200 mm long and 20 mm wide at the base. The margins of the leaves are armed with very small brownish teeth. The inflorescences are fairly short, up to 200 mm long, with comparatively large, swollen yellow flowers.

The species flowers from June to August.

This species is doubtfully distinct from *A. dichotoma* but can generally be recognised by the absence of a main trunk. It has similar inflorescences and flowers.

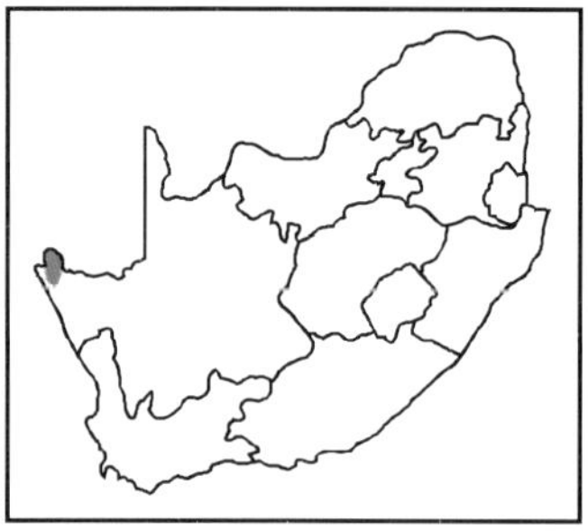

Aloe ramosissima is restricted to the Richtersveld and southern Namibia.

The species is regarded as vulnerable as a result of mining activities and overgrazing.

Due to the smaller stature of *A. ramosissima*, its common name is Maiden's quiver tree or Nooienskokerboom. *Ramosissima* means 'very much branched'. The national tree number is 30.2.

Ben-Erik van Wyk

Pitta Joffe – NBI

Single-stemmed aloes

Group 2

This group of aloes is easily recognised by the **single, erect main stem** which is sometimes branched only in the upper part. There is usually only a **single large rosette of leaves** (or several large rosettes if the stem is branched). The **trunk** is almost invariably **single at the base** and clothed **with dry leaf remains**. In some species with single rosettes, damage to the growing tips may result in multiple stems. In other species, one to several side-branches may be present in old specimens. All the species are spectacular when in flower, due to the large, densely flowered and brightly coloured racemes. The single-stemmed aloes are widely distributed in South Africa and are commonly planted in gardens and parks.

SPECIES

A. africana
A. alooides
A. angelica
A. comosa
A. excelsa
A. ferox
A. lineata
A. littoralis
A. marlothii
A. pluridens
A. pretoriensis
A. rupestris
A. speciosa
A. thraskii
A. vryheidensis

Pitta Joffe – NBI

Aloe marlothii

Aloe africana

This aloe is a single-stemmed plant of 2 to 4 m tall, with dry leaf remains persisting on the stem. The spreading and recurved leaves are dull green to greyish-green, with reddish spines along the edges. There are sometimes a few spines on the upper and lower surfaces as well. The inflorescences are single or more often have two to four strongly tapering racemes. The flowers are bright yellow-orange, with the open flowers characteristically curved upwards.

Flowering occurs from July to September, but it is not unusual to find one or two plants in flower at other times of the year.

Aloe africana can easily be distinguished from similar single-stemmed species by the more slender and more spreading to recurved leaves. The strongly tapering inflorescences and the peculiar upcurved flowers (similar to those of the dwarf aloe *A. longistyla*) are very distinctive.

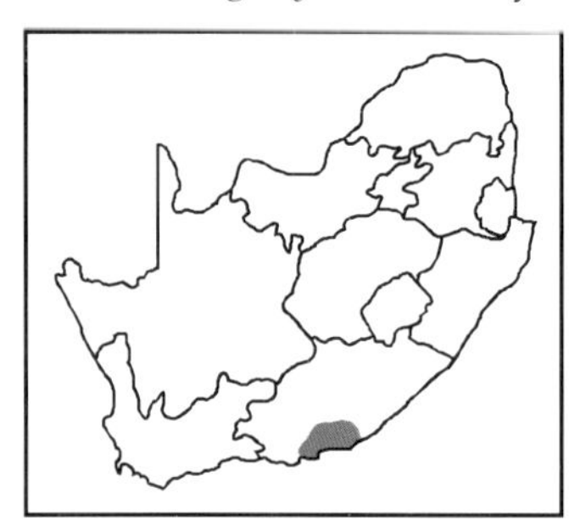

The distribution area extends from near the Gamtoos River eastwards to Port Alfred. It is particularly common in the Port Elizabeth and Uitenhage districts. *Aloe africana* is usually found in dense bushveld vegetation and rarely in open veld (the opposite is true for *A. ferox*).

Aloe africana is not threatened.

The attractive flowers and gracefully recurved leaves make *A. africana* a spectacular garden plant in frost-free areas. The common name is Uitenhage aloe or Uitenhaagsaalwyn. *Africana* means 'from Africa'. The national tree number is 28.2.

Frits van Oudtshoorn

Gideon Smith

Geoff Tribe

Aloe alooides

Plants are single-stemmed and up to 2 m tall, with the upper part of the trunk covered with old dry leaf remains. The broad leaves are green or sometimes slightly reddish and curve down strongly, so that the tips almost touch the stem. The leaf margins have small spines along the edges, but the upper and lower surfaces are smooth. The inflorescence is always single, but up to five may be present on each plant. It is long, slender and densely covered with large numbers of greenish-yellow flowers. The flowers are very small (only up to 10 mm long), bell-shaped (not tubular) and without a stalk.

The species flowers in July and August.

The recurved, almost thornless leaves of *A. alooides* are very similar to those of *A. thraskii*, but the latter is a coastal species with a much-branched inflorescence of short, broad racemes. *Aloe alooides* may also be confused with the closely related *A. spicata*, but this species forms dense clumps of short-stemmed plants. *Aloe vryheidensis* is another close relative, although it has erect leaves and much shorter racemes.

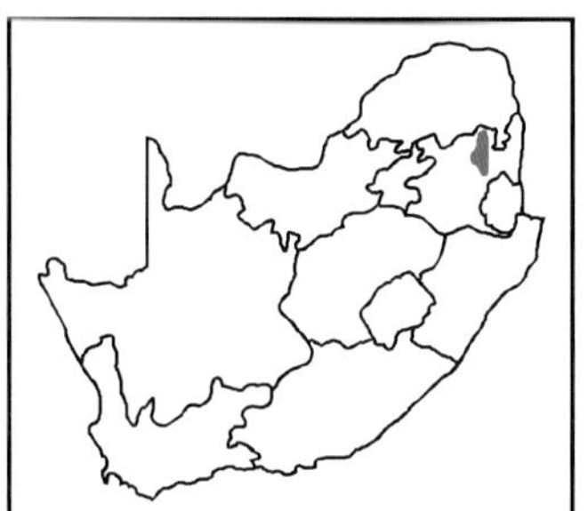

Aloe alooides has a rather restricted distribution in the mountains of Mpumalanga. The plants grow in shallow soil on open dolomite ridges.

Relatively large numbers of plants occur in some inaccessible places and the species is not threatened.

Aloe alooides is a very attractive aloe; the drooping, recurved leaves and golden, snake-like inflorescences result in an unusual and distinctive appearance. It makes a spectacular garden plant, provided that it is planted in a warm spot with well-drained soil. This aloe was originally thought to be an aloe-like species of *Urginia*, hence the scientific name *alooides* which means 'resembling an aloe'. The common name is Graskop aloe or Graskopaalwyn. The national tree number is 28.3.

Ben-Erik van Wyk

Ben-Erik van Wyk

Aloe angelica

This species is usually single-stemmed and up to 4 m high, but some individuals may be branched. The stems are long and slender, with dry leaves covering only the upper half. The leaves at the top of the rosette spread out horizontally, while the lower ones are strongly recurved. Small sharp teeth occur along the leaf margins, but the upper and lower surfaces have no spines or prickles. The inflorescence is much-branched into ten to twenty racemes on long, erect branches. Each raceme is short and rounded, with a distinctive bicoloured appearance. This colour effect results from the reddish flower buds which turn greenish-yellow when they open. The flowers are short and broad, but distinctly tubular.

Flowering occurs in June.

Aloe angelica may be confused with *A. alooides* or *A. thraskii* when not in flower, but the lower half of the stem becomes smooth without dry leaf remains. The short, rounded and bicoloured racemes clearly separate it from all other superficially similar species such as *A. pluridens*.

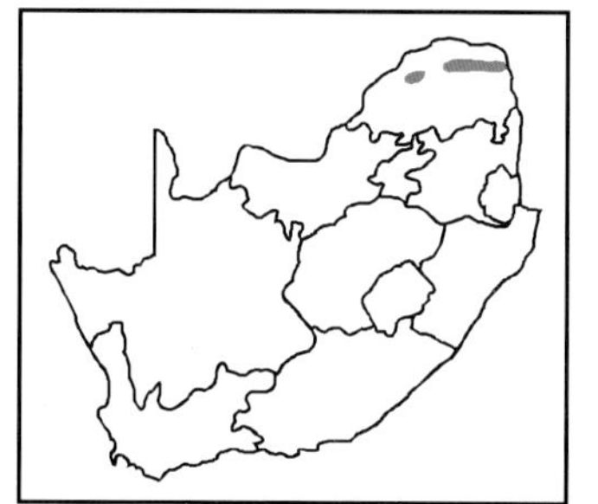

The distribution area is rather restricted and the species is limited to the Soutpansberg and Blouberg in the Northern Province. It grows in open places or in dense bush.

Aloe angelica is locally common and is not threatened.

This aloe has no close relatives but the chemistry of the leaf sap suggests a relationship with *A. arborescens*. Plants do not grow well outside their natural habitat and are easily killed by frost. The common name is Wylliespoort aloe or Wylliespoortaalwyn. The national tree number is 28.4.

Eben van Wyk

Eben van Wyk

Eben van Wyk

Aloe comosa

This striking aloe is single-stemmed and 2 m high, but when in flower reaches a total height of 5 m. The gracefully recurved leaves are distinctly grey and about 700 mm long. The pinkish margins are armed with small reddish-brown teeth. The erect, very tall inflorescences of up to 3 m long, are usually unbranched but may sometimes have as many as five branches near the base. The flowers are red in the bud stage, but open to a dull pink or whitish colour, resulting in an attractively contrasting bicoloured raceme.

The species flowers in December and January.

An exceptional feature of *A. comosa* is the very tall and slender inflorescence. The buds are upright and tightly pressed against the peduncle. This characteristic, together with the large and overlapping bracts, further enhances the slender appearance of the racemes. The buds are bright red and turn whitish upon opening, at which point they become pendulous and closely pressed against the flowering stem, but in the opposite direction.

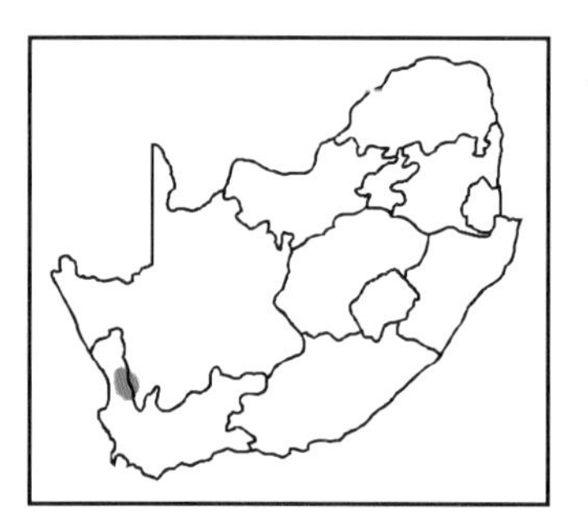

Aloe comosa is restricted to a small region just north of Clanwilliam in the Western Cape Province.

The species is regarded as rare due to its small world population, illegal plant collection and unsound agricultural practices, such as overgrazing.

In frost-free areas, *A. comosa* will make a beautiful feature in the garden. However, this species should be watered sparingly in the summer since it occurs naturally in a very dry part of the winter rainfall area of the Cape. *Comosa* means 'bearing a tuft of leaves'. The common name is Clanwilliam aloe or Clanwilliam-aalwyn, and the national tree number is 28.7.

Ben-Erik van Wyk

Ben-Erik van Wyk

Aloe excelsa

The plants have simple stems of 2 to 4 m high which are densely covered with dry leaf remains. The leaves are dull green to reddish-green, with reddish-brown marginal teeth along the edges. The upper surfaces are generally smooth, but the lower surfaces usually have scattered prickles, particularly in young plants. The inflorescence is branched into ten to fifteen racemes, of which the upper ones are erect but the lower ones are slightly oblique (sloping). Each raceme is short, cylindrical and densely flowered. The flowers are tubular, about 30 mm long, and dark red or orange-red in colour.

Flowering occurs in August and September.

Aloe excelsa is closely related to *A. rupestris* but is much less common than the latter. It may be distinguished by the oblique outer racemes and the darker colour of the flowers. The leaves are also larger and more spiny.

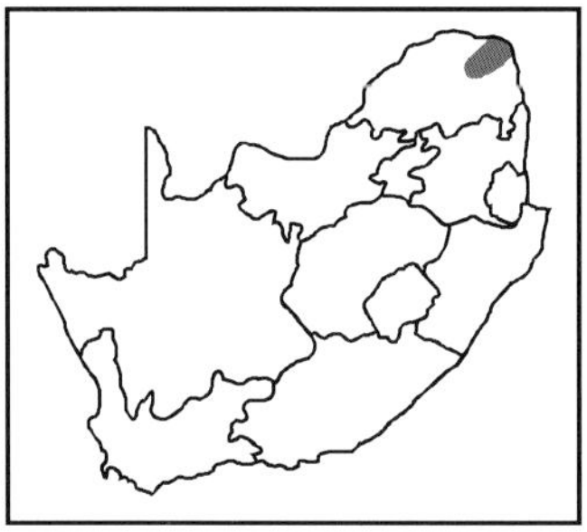

In South Africa, this species is restricted to the extreme north-western corner of the country and is only found near the Limpopo River. It also occurs in warm dry valleys and hills of Mozambique, Zimbabwe, Zambia and Malawi.

Aloe excelsa is very localised in South Africa, but is widely distributed further north. It is not threatened.

The neat appearance and brightly coloured flowers of this species make it an attractive garden plant. In warm areas, it grows and flowers well in cultivation. *Excelsa* means 'lofty' or 'high'. The common name is Zimbabwe aloe or Zimbabwe-aalwyn, and the national tree number is 28.8.

Darryl Plowes

Gideon Smith

Pitta Joffe – NBI

Aloe ferox

A robust single-stemmed plant, usually 2 m high but up to 5 m in old specimens. The leaves are broad and dull green to greyish-green, but turn reddish in colour when under drought stress. Dark brown spines are present along the edges and sometimes also on the upper and especially the lower surfaces of the leaves. The dry leaves are persistent on the lower parts of the stem. The inflorescence is branched into five to twelve erect racemes. The flowers are usually bright orange-red, but bright red, yellowish and even white forms are found. The tips of the inner perianth segments are usually dark brown but may be white in KwaZulu-Natal and adjacent areas of the Eastern Cape Province.

Flowering occurs from May to August in most of the distribution area, but somewhat later (September to November) in the colder northern parts.

Aloe ferox is similar to the KwaZulu-Natal form of *A. marlothii* (previously known as *A. spectabilis*), but the latter usually has the individual racemes at a slight angle and not erect as in *A. ferox*. Also, the tips of the inner perianth segments of the KwaZulu-Natal form of *A. ferox* are white, not dark brown as in the KwaZulu-Natal form of *A. marlothii*. Furthermore, the leaf surfaces of *A. ferox* in this region are virtually devoid of spines, but are distinctly spiny in *A. marlothii*. Elsewhere in the distribution area, *A. ferox* is unlikely to be confused with other species.

This is one of the most widely distributed species, occurring from the Swellendam area in the west, to the dry parts of the Western and Eastern Cape Provinces and southern KwaZulu-Natal, with a few localities in south-western Lesotho and the extreme south-eastern part of the Free State. It occurs in a wide range of habitats, on mountain slopes, rocky places and flat open areas. The species shows remarkable adaptability in terms of rainfall, and flourishes in extremely dry areas of the Karoo but also in relatively wet parts of the eastern part of its distribution range.

Aloe ferox is a very common species and is not threatened. However, in the Free State and KwaZulu-Natal it is protected by legislation. Overexploitation or careless harvesting practices (i.e. the removal of too many leaves per plant) may lead to local extinction at some localities.

As a medicinal plant of commercial importance, *A. ferox* has no rival in Southern Africa. For more than two hundred years the golden-brown leaf exudate has been used for the production of the purgative drug known as Cape Aloes. This species has numerous common names, of which Bitter aloe is the most commonly used. *Ferox* means ferocious, referring to the spiny leaves. The national tree number is 29.2. *Aloe ferox* is commonly grown in gardens, and we recommend the KwaZulu-Natal form (previously known as *A. candelabrum*) for cultivation. It has gracefully recurved leaves and bright coral-red flowers.

Pitta Joffe – NBI

Cameron McMaster

Ben-Erik van Wyk

Ben-Erik van Wyk

Ben-Erik van Wyk

Aloe lineata

Plants are invariably stemmed. The stems, which may reach a height of about 2 m, are covered with the remains of dried leaves. The rosettes are medium-sized and quite compact. The bluish-green leaves have distinct reddish longitudinal lines on both surfaces and the margins are armed with firm, pungent, reddish-brown teeth. Up to four simple racemes are produced consecutively from a single rosette. The flowers are fairly large and are covered by large bracts in the bud stage. Flower colour ranges from light pink to bright red.

Aloe lineata flowers in summer, from January to March, whereas the var. *muirii* flowers from June to November, with a peak from July to September.

The species is primarily distinguished by the bluish leaves with distinct reddish lines. Other useful characteristics include the flowering time, medium size and presence of a stem. In addition, its rosettes are not as robust as those of the more or less winter-flowering *A. ferox* and *A. africana* with which it co-occurs over much of the eastern parts of its distribution range.

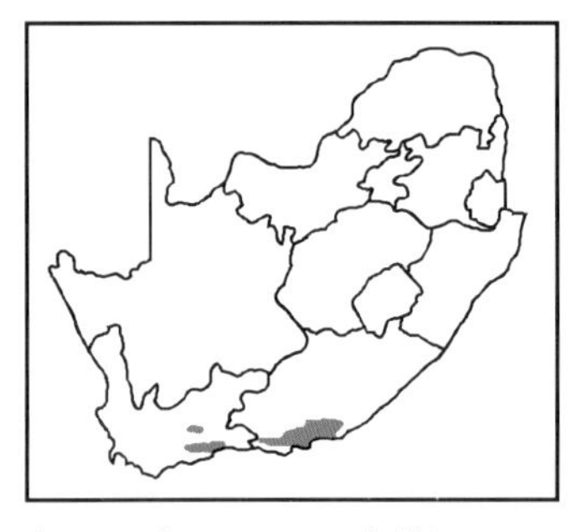

Aloe lineata is a component of thicket and dense bushy vegetation. Its distribution stretches from Riversdale in the west to Grahamstown in the east.

The species is not threatened.

Aloe lineata is a most rewarding and spectacular garden plant. It is not difficult to cultivate, except where frosts are severe. In addition to the typical form, the var. *muirii* is recognised. It has yellow-green leaves with very distinct red striations and larger marginal teeth. *Lineata* means 'marked with fine parallel lines'.

Kobus Venter

George Lubbers

Forsyth van Nierop

Aloe littoralis

Plants are single-stemmed and up to 3 m high. The greyish-green leaves are about 600 mm long and 120 mm wide at the base, with numerous small white spots on the upper surfaces and especially the lower surfaces. The sharp teeth along the leaf margins are brown to reddish-brown, but each one arises from a white base. The inflorescences are much-branched into ten or more erect, narrow, oblong, somewhat sparsely flowered racemes. The flowers are tubular, up to 30 mm long and vary from pink to red, often with a greyish waxy layer which gives them a silvery appearance. The open flowers become distinctly yellowish towards the mouth.

In South Africa, flowering occurs mainly in February and March.

Aloe littoralis is a variable species, but the South African form is without exception tall and single-stemmed. It may be distinguished from other single-stemmed aloes by the numerous small spots on the upper and particularly the lower leaf surfaces. Young plants are more copiously spotted than large specimens. The flower colour (usually silvery pink with yellow at the mouth) is also a useful diagnostic characteristic.

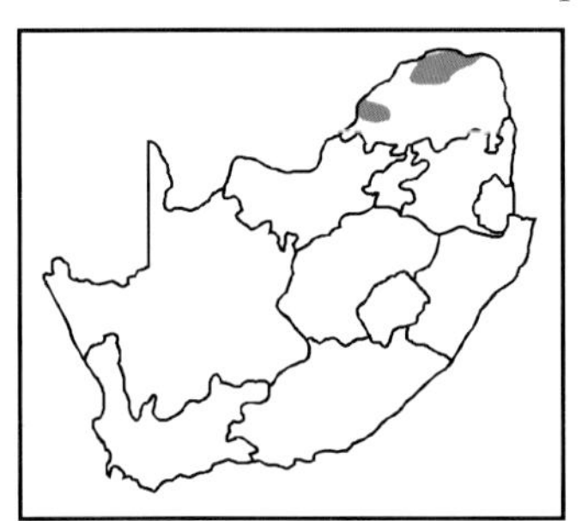

The species is widely distributed in Southern Africa and also occurs in Angola, Namibia, Botswana, Zimbabwe and Mozambique. In South Africa, it is found only in the extreme northern and western parts of the Northern Province, where it grows in dry open bushveld vegetation.

Aloe littoralis is a widely distributed and common species, so that it is not threatened. In South Africa, however, it is less abundant and should be protected.

This attractive aloe grows well in cultivation and even young plants will produce flowers. Common names include Mopane aloe, Mopanie-aalwyn and Bergaalwyn. The scientific name *littoralis*, which means 'growing on the sea shore', is perhaps suitable to describe the place where it was originally found in Angola, but elsewhere the name is not very appropriate. The national tree number is 29.4.

Geoff Nichols

Frits van Oudtshoorn

Eben van Wyk

Aloe marlothii

Plants are large and invariably single-stemmed, usually 2 to 4 m high but occasionally up to 6 m or more. The leaves are broad, dull green to greyish-green, with dark brown spines along the margins and usually also on the upper and especially the lower surfaces. The dry leaves are persistent on the lower parts of the stem. The inflorescence is much-branched into about twenty to thirty racemes. The racemes are usually distinctly horizontal, but are slanted to almost erect in the southern form from KwaZulu-Natal. The flowers are usually bright orange-red, but can vary from bright red to yellow.

The flowering time is from May to September.

Aloe marlothii can easily be distinguished from all other single-stemmed aloes by the horizontal or at least slanted racemes with erect flowers. The KwaZulu-Natal form of the species (previously known as *A. spectabilis*) is similar to *A. ferox*, but the racemes of the latter are invariably upright and not at a slight angle. (Other differences are given under *A. ferox*.)

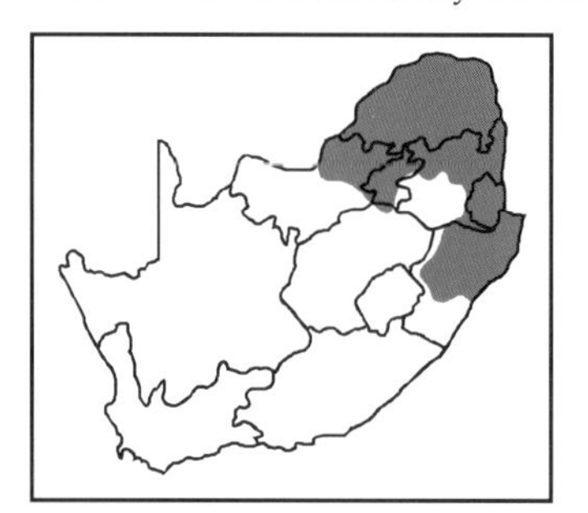

Aloe marlothii is a typical feature of the landscape of a large part of Southern Africa. It occurs from KwaZulu-Natal in a broad band westwards to Botswana and northwards to Mozambique and Zimbabwe. Plants grow in a variety of habitats, from open bushveld to exposed rocky places.

Aloe marlothii occurs in vast numbers and is not threatened.

The KwaZulu-Natal form of the species merges gradually with the typical forms, and even the leaf sap chemistry shows no differences. In the past, the species has been a source of drug aloes, but it is chemically very variable and is no longer commercially used. There are numerous folk uses, however, and the dry leaves are commonly ground and mixed with snuff. There are several common names, such as Mountain aloe, Bergaalwyn, Boomaalwyn, umHlaba or imiHlaba (Zulu) and Kgopha (Sotho). The scientific name commemorates the famous botanist, H.W. Rudolf Marloth. The species is often grown in gardens and can withstand quite severe frost. The red form from the Utrecht district of KwaZulu-Natal is very striking, although the rich golden colour of the typical form is equally attractive. The national tree number is 29.5.

Frits van Oudtshoorn

Eben van Wyk

Frits van Oudtshoorn

Gerhard Steyn

Peter Winter

Aloe pluridens

Plants are single-stemmed or form several tall branches with gracefully recurved leaves in large rosettes. Sometimes numerous small plantlets are found on the otherwise smooth stems. The leaves are bright green, distinctly fleshy and armed with soft, small white teeth. The remains of the dried leaves are papery in texture and are restricted to the upper parts of the stems. The inflorescences have up to four branches with long conical racemes protruding well above the rosettes. The flowers are usually orange or pinkish-red, but a yellow form is also known.

Flowering occurs from May to July.

At a distance the stems and terminal rosettes of *A. pluridens* look like those of *A. angelica*. However, the latter species is restricted to the Northern Province and has very different, shorter, more head-shaped racemes. *Aloe pluridens* is also similar to *A. arborescens* but can be distinguished by being taller and more slender. The leaf sap is watery with a distinct odour, and not yellowish as in *A. arborescens*.

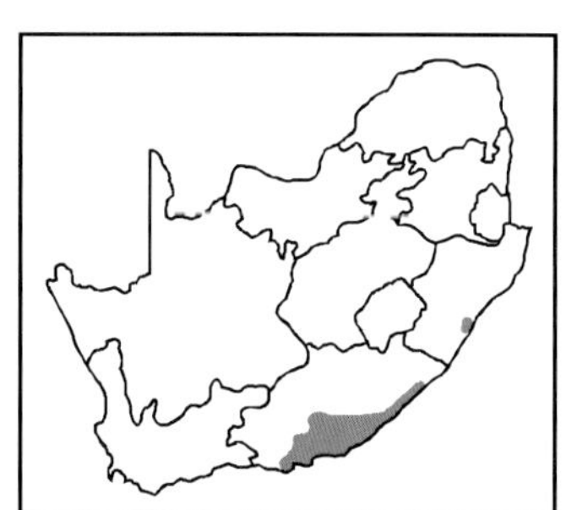

The species occurs in a broad coastal band from Humansdorp in the Eastern Cape Province eastwards to Durban in KwaZulu-Natal. It is usually associated with impenetrable thicket vegetation.

Aloe pluridens is not threatened.

This aloe makes a beautiful garden plant but needs protection against frost. It appears to be less prone to aloe snout beetle infestations than most other species. The scientific name *pluridens* refers to the many marginal teeth of the leaves (*pluri* means 'many'; *dens* means 'teeth'). The common name is French aloe or Fransaalwyn. The national tree number is 30.1.

Ben-Erik van Wyk

Ben-Erik van Wyk

Pitta Joffe NBI

Aloe pretoriensis

Plants are generally medium-sized and solitary. Old specimens have more or less leafless stems of up to 1 m tall. The erect, light bluish-green leaves form neat rosettes. The leaves are indistinctly striped and die back from the tips. The inflorescences are very tall and branch from above the middle. The racemes are distinctly conical and carry long, pencil-shaped flowers which are constricted near the mouth. The flower colour is usually bright red, but orange-red forms are sometimes encountered.

Aloe pretoriensis flowers from May to July.

This is a distinctive species which has exceptionally tall inflorescences for such a relatively small plant. Furthermore, the dried leaf tips take on the same bright red colour as the flowers. Young plants may be misidentified as *A. suprafoliata*. (Differences between these two species are discussed under the latter.)

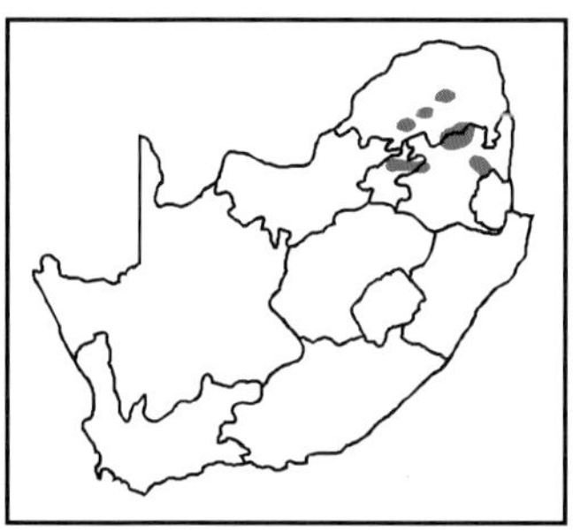

The species occurs in exposed positions in grassland in northern Gauteng, the Northern Province, Mpumalanga, Swaziland and Zimbabwe.

The species is not threatened.

Aloe pretoriensis does well in cultivation in the summer rainfall region of Southern Africa, provided that it is planted in a frost-free spot. This is one of the most beautiful of all the species of *Aloe*. *Pretoriensis* means 'from Pretoria', although the species has a much wider distribution.

Pitta Joffe – NBI

Pitta Joffe – NBI

Eben van Wyk

Aloe rupestris

Plants usually have simple stems of up to 8 m high, with only the upper part covered with dry leaf remains. The leaves are erect and spreading, with reddish-brown marginal teeth. The upper and lower surfaces are smooth and without prickles. The inflorescence is much-branched, with about fifteen more or less erect racemes. Each of the densely flowered racemes is short and cylindrical, and rather broad. The flowers are tubular and vary from yellow to bright orange, while the protruding stamens are dark red. This results in an attractive bicoloured effect.

The plants flower in August and September.

Aloe rupestris is closely related to *A. excelsa* and *A. thraskii* but can readily be distinguished by its general appearance (apart from details of the leaves, inflorescences and flowers that are given under the latter two species).

This species is relatively common in KwaZulu-Natal, Swaziland and southern Mozambique. The plants grow in hot valleys and are usually found in groups amongst trees, so that they are less conspicuous than other species which grow in the open.

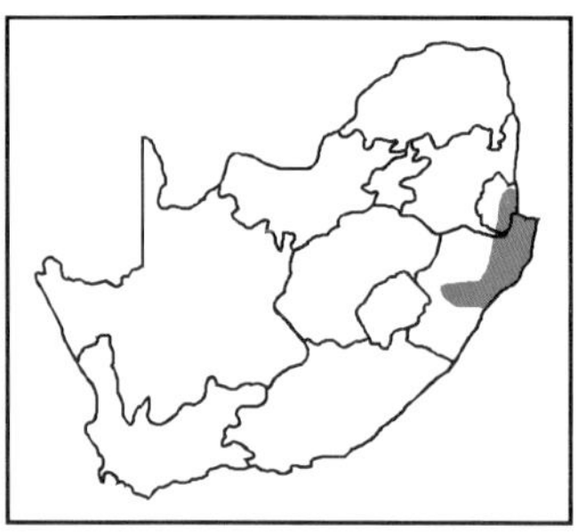

Aloe rupestris is widely distributed and is not threatened.

As a result of its attractive flowers and erect growth form, this species has become a popular garden plant all over South Africa. It is unfortunately very susceptible to white scale. Plants withstand moderate frost if placed in a protected spot. The Zulu common names are inKalane, umHlabanhazi or uPhondonde, and the species is sometimes known in Afrikaans as Kraalaalwyn. *Rupestris* means 'growing in rocky places'. The national tree number is 30.3.

Ernst van Jaarsveld - NBI

Ben-Erik van Wyk

Trevor Coleman

Aloe speciosa

This aloe is single-stemmed or branched at various heights from the ground and is 3 to 6 m high at maturity. The rosettes are invariably tilted to one side, with slender, blue-green leaves of up to 900 mm long. The leaves have narrow, pinkish edges and are armed with minute, harmless teeth. The inflorescence is borne on a short peduncle, giving the impression that the racemes are carried within the rosettes. The racemes are short and densely flowered, and about 300 mm long. They are invariably unbranched, but up to four of them may be present in each rosette. When fully open, the flowers change colour from red to greenish-white.

This species flowers from July to September.

Aloe speciosa is rarely found as solitary plants and usually occurs in very large numbers, often covering the landscape as far as the eye can see. The leaves are irregularly arranged, giving the rosettes an untidy appearance. As in the case of *A. comosa*, the buds are red and the flowers turn whitish when they open. However, *A. speciosa* can be distinguished by its much shorter and thicker racemes and distinct geographical distribution.

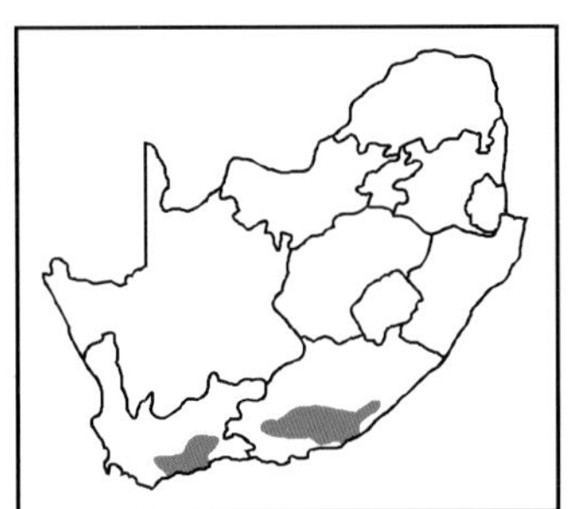

This species occurs from Swellendam in the west in a broad band eastwards to the Kei River. It is particularly abundant in the Eastern Cape Province, for example between Uitenhage and Jansenville.

Aloe speciosa is exceptionally common and is not threatened.

This is undoubtedly one of the most beautiful aloes for the garden, and will grow well in most areas, even where moderate frost occurs. In addition, it appears to be unaffected by aloe snout beetles, perhaps due to its peculiar leaf sap chemistry. The common names are Tilt-head aloe, Slaphoringaalwyn and Spaansaalwyn. *Speciosa* means 'showy'. The national tree number is 30.5.

Frits van Oudtshoorn

Ben-Erik van Wyk

Leo Thamm

Aloe thraskii

A robust, single-stemmed plant, usually 2 m high but sometimes up to 4 m, with dry leaves persisting on the stem. The leaves are broad, deeply U-shaped in cross-section, dull green to greyish-green and strongly curved downwards, so that the tips touch the stem. Small reddish spines are present along the edges and sometimes also in a line along the middle of the lower surfaces. The inflorescences are branched, each with fifteen to twenty-five erect racemes. Young plants produce only one inflorescence, but old specimens may have three or four. Each raceme is broad, cylindrical and densely flowered. The flowers are yellowish-orange, with the protruding parts of the stamens bright orange.

Flowering occurs in June and July.

Aloe thraskii is closely related to *A. excelsa* and *A. rupestris,* but can be distinguished from both these species by the strongly recurved leaves. In this respect it closely resembles *A. alooides,* although the latter has long and slender racemes.

This strikingly beautiful species is widely distributed along the east coast of South Africa and is always found on sand dunes near the beach. It usually grows in dense coastal bush and not in open exposed places.

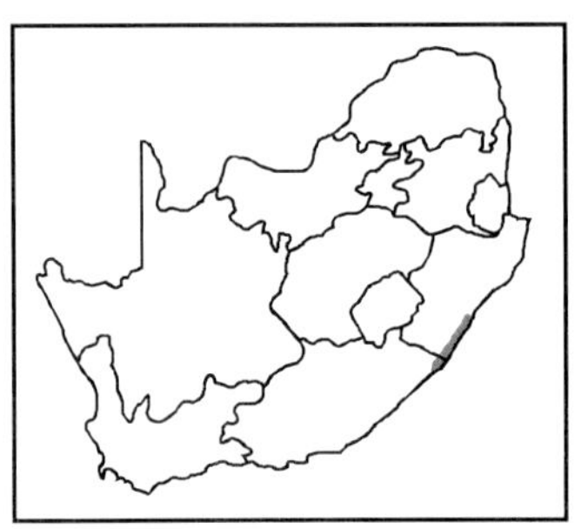

Aloe thraskii is not threatened, but urban development along the coast has resulted in extinction at some localities.

As the common names Strand aloe or Dune aloe (Strandaalwyn) suggest, this aloe is a good choice for gardens along the coast and deserves to be planted more often. Few other species can rival its graceful beauty. The species is named after a long forgotten person by the name of Thrask. The national tree number is 30.7.

Duncan Butchart

Leo Thamm

Pitta Joffe NBI

Aloe vryheidensis

Plants are single-stemmed and up to 2 m high, usually with a short trunk covered with old dry leaf remains. The stems of old specimens appear to be strongly tapering downwards as a result of the way in which the leaf remains gradually weather away from the bottom. The leaves are greyish-green but often turn reddish in winter. They are erect or spreading (not recurved) and have small, sharp teeth along the margins only. The racemes are invariably single, but up to five may be formed on each plant. They are erect, narrow, cylindrical and densely multiflowered. The distinct bottle-brush appearance is due to the small, densely crowded bell-shaped flowers with their exserted stamens.

The species flowers in July and August.

Aloe vryheidensis is a variable species and some forms are very similar to *A. spicata*, which has almost identical inflorescences. It may be confused with single-stemmed specimens of the latter, but can be distinguished by the shorter, thicker, more erect leaves and the more compact rosettes.

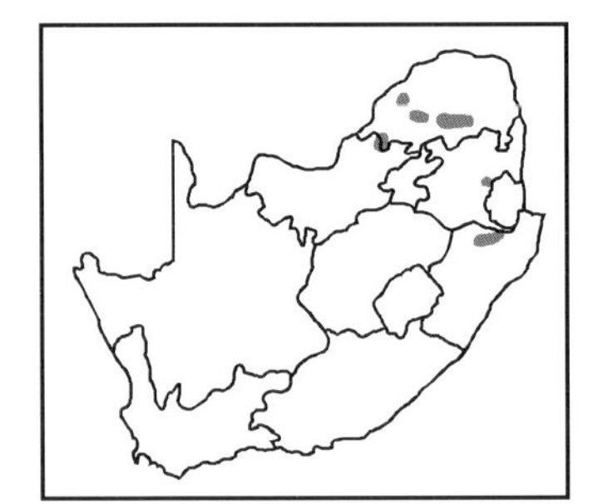

The geographical distribution area is from the mountainous parts of northern KwaZulu-Natal northwards to the Wolkberg and Strydpoortberge in the Northern Province. Plants grow in exposed rocky places, often on dolomite. In the northern part of the distribution area, the plants are generally taller than in the south, with a more pronounced trunk.

Large numbers of plants are found at several localities and the species is not threatened.

Aloe vryheidensis is an attractive garden plant and grows well in alkaline soil. The common name Bruinaalwyn has been recorded. Our concept of *A. vryheidensis* includes the species formerly known as *A. dolomitica*, which was said to differ mainly in the slightly longer stems. The species was named after the northern KwaZulu-Natal town of Vryheid. The national tree number is 29.1.

Ben-Erik van Wyk

Ben-Erik van Wyk

Ben-Erik van Wyk

Multistemmed aloes

Group 3

This group is a conglomerate of **medium to large shrubs** that are usually **multistemmed** at or near ground level. In most species, the leaves are strongly recurved and are soft in texture, with rather harmless marginal teeth. In *A. claviflora* and *A. falcata* however, the leaves are curved upwards and are very firm in texture, with sharp dark brown or black marginal teeth. The group includes some of the most widely cultivated and popular garden subjects.

SPECIES

A. arborescens
A. castanea
A. claviflora
A. falcata
A. hardyii
A. mutabilis
A. spicata
A. succotrina
A. vanbalenii

Gideon Smith

Aloe arborescens

Aloe arborescens

Plants form large, many-branched shrubs or trees of about 2 m high with the leaves borne in apical rosettes. The greyish-green to bright green leaves vary considerably in length, averaging from 500 to 600 mm. The leaves are sickle-shaped and strongly recurved to flatly spreading. The leaf margins are commonly armed with firm, white teeth or teeth which are the same colour as the leaves. A form with smooth margins is sometimes encountered. The inflorescence is up to 900 mm long and is usually simple, but occasionally has up to two side-branches. The raceme is conical in shape and bears scarlet, orange, pink or yellow flowers.

The species flowers from May to July.

Aloe arborescens is most closely related to *A. mutabilis* which can generally be distinguished by its bicoloured racemes and broader leaves. *Aloe mutabilis* is a smaller plant with shorter, more or less trailing stems and it has a restricted distribution from the Witwatersrand northwards to Pietersburg.

The species occurs from the Cape Peninsula, along the eastern coast of Southern Africa, through KwaZulu-Natal, Mpumalanga and the Northern Province, to Mozambique, Zimbabwe and Malawi. This is the third widest distribution of all the species of *Aloe*. It grows mainly in mountainous areas, occasionally in dense bush, or on exposed ridges or krantzes.

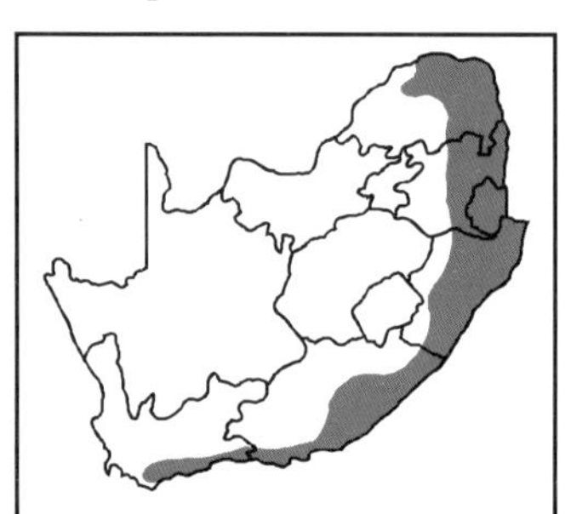

Aloe arborescens is exceptionally common and is not threatened.

Aloe arborescens can be easily propagated from cuttings or truncheons and is probably the most widely cultivated species of the genus in the world. In South Africa it is commonly used as a garden ornamental and also as a hedge plant, to protect agricultural fields or stock. Even today cuttings of the species are sold in muthi shops as barrier plants. In the Orient, it is commonly grown in domestic gardens as a convenient first-aid treatment for burn wounds and abrasions. Extracts of the leaves have also been used to treat X-ray burns. *Aloe arborescens* hybridises readily with other species of *Aloe* with which it co-occurs. Particularly striking crosses between *A. arborescens* and *A. ferox* occur around Mossel Bay and on the banks of the Gourits River in the southern Cape. The common names of the species are Krantz aloe and Kransaalwyn, or Inkalane in Zulu. *Arborescens* refers to the tree-like habit. Its national tree list number is 28.1.

Forsyth van Nierop

Ben-Erik van Wyk

Ben-Erik van Wyk

Aloe castanea

Plants are usually small trees with a single main trunk at ground level and several spreading branches higher up. The stems are covered with the remains of old dried leaves for quite a distance, with only the bottom-most parts being exposed. The firmly textured leaves are up to 1 m long, and narrow and oblong. They are arranged in rather untidy, terminal rosettes. The leaf margins are armed with firm, small, brown teeth. The distinctive inflorescences are not neatly erect as in most other aloes, but are curled and snake-like in appearance. The tiny dark orange-brown flowers with their prominent stamens are densely packed and give the long thin racemes a brush-like appearance.

The species flowers from June to August.

This is a very distinctive aloe with its spreading branches and snake-like inflorescences. In addition, the dark brown nectar is a characteristic shared by few other aloes.

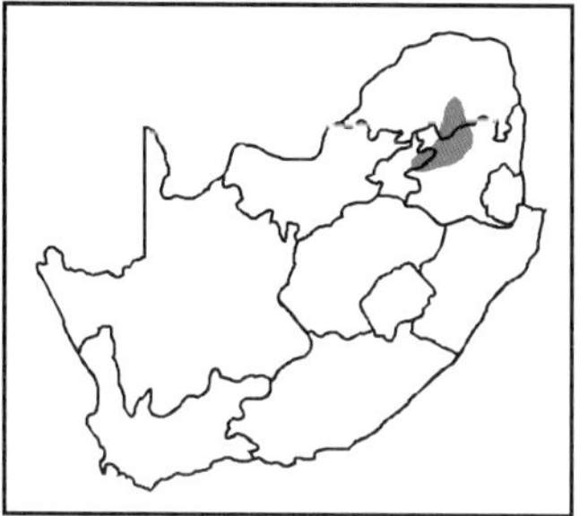

Aloe castanea occurs in a broad band from Witbank northwards to Lydenburg in Mpumalanga and Pietersburg in the Northern Province.

The species is exceedingly abundant and is not threatened.

Aloe castanea is an attractive garden plant that warrants wider cultivation. It will tolerate extreme conditions and grows particularly well on the Witwatersrand. *Castanea* is the Latin genus name for the chestnut, and it was used for this aloe to describe the chestnut-brown colour of the nectar. The common name is Cat's-tail aloe or Katstertaalwyn, alluding to the appearance of the inflorescences. The national tree number is 28.6.

Ben-Erik van Wyk

Ben-Erik van Wyk

Aloe claviflora

Plants grow in dense groups or circular patches of ten or more heads. Stems, if present, are short and grow horizontally along the ground. The rosettes are not erect but face outwards, giving them a characteristically asymmetrical shape. The leaves are greyish-green and firm in texture, up to 200 mm long, and have sharp brown spines along the margins. A few spines are also present along the middle of the lower surface, but only towards the tip of the leaf. A peculiar feature is the angle at which inflorescences are produced — never erect but always at a slanted angle, often almost horizontal. One or two appear from each rosette. They are usually unbranched, but may rarely have up to four branches. The racemes are oblong, densely flowered and up to 300 mm long. Each flower is distinctly club-shaped, with the widest part near the mouth and the tube strongly tapering and gradually merging with the flower stalk. The buds and young flowers are bright red, but turn yellow or whitish with age, so that the racemes have an attractive bicoloured appearance.

Flowering occurs in August and September.

The slanted inflorescences and club-shaped flowers are unique to this species. Even when not in flower, *A. claviflora* may be identified by the peculiar way in which the dense groups of rosettes gradually open up in the middle. The old rosettes in the centre die back, leaving a circular patch of young plants. Other useful diagnostic features include the firm, rough leaf surfaces and dark brown marginal thorns.

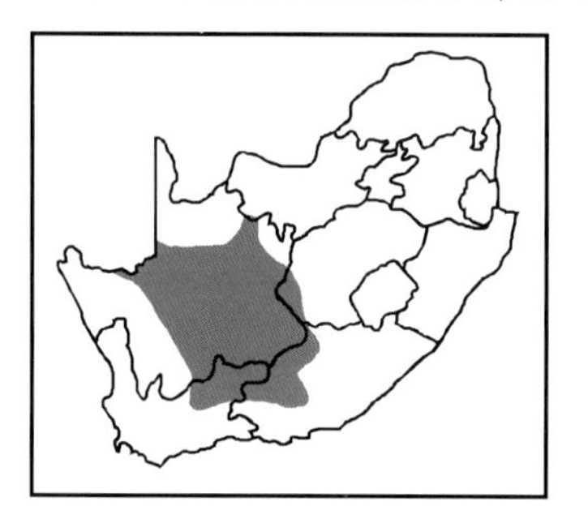

Aloe claviflora is widely distributed in the dry interior of South Africa, where it grows in well-drained places on rocky slopes or flat stony areas.

The species is very common and is not threatened.

Two unique features of this strikingly beautiful aloe are reflected in the Afrikaans common names. Kanonaalwyn (meaning 'cannon aloe') refers to the oblique angle of the inflorescence, while Kraalaalwyn refers to the peculiar growth habit. The scientific name *claviflora* means 'club-shaped flowers'.

Geoff Tribe

Ted Scholes

Leo Thamm

Aloe falcata

Plants usually grow in dense groups, with the rosettes pointing outwards and often almost lying on their sides, with the leaves curved upwards. Stems are short or absent. The leaves are green to greyish-green, firm in texture, with rough, sandpaper-like surfaces. The leaves have pale to dark brown teeth along the margins. The inflorescences are branched into several (up to ten) racemes, and they are erect and not at an oblique angle. The racemes are long and sparsely flowered, usually dull red or rarely yellow. The flowers are tubular, rather long and narrow, about 40 mm long and 6 mm in diameter.

The species flowers in December.

Aloe falcata is a close relative of *A. claviflora* and also has asymmetrical rosettes and firm, rough-textured leaves. Despite the resemblance, it may easily be distinguished by the many-branched, erect inflorescences and tubular (not club-shaped) flowers. The plants also form dense groups, but these do not open up in the middle as in *A. claviflora*.

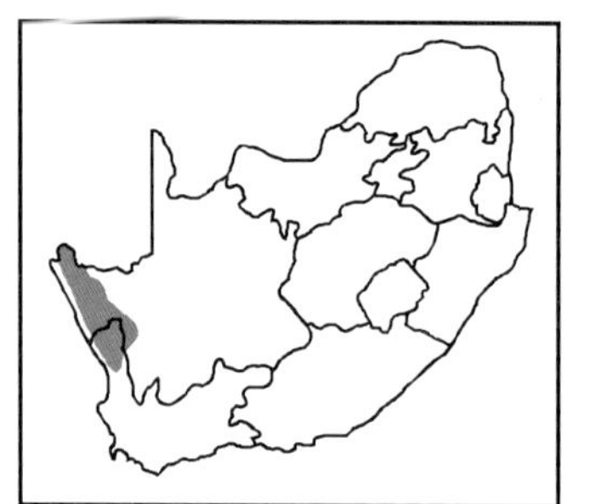

The species grows on dry, flat sandy places in the north-western parts of the Cape, from the Knersvlakte (the southern limit near Klawer) northwards to the Richtersveld.

Aloe falcata is considered vulnerable as a result of illegal collection and agricultural activities.

The Latin word *falcata* refers to the sickle-shaped or falcate leaves which curve upwards from the strongly tilted rosettes. *Aloe falcata* is a spectacular species but it does not thrive outside the natural habitat.

Ben-Erik van Wyk

Frits van Oudtshoorn

Ben-Erik van Wyk

Aloe hardyi

The species has hanging branches and usually grows on rocky cliffs. The dense rosettes have more or less pendulous leaves reaching a length of more than 700 mm. Soft, yellowish teeth occur along the leaf margins. The inflorescences are invariably unbranched and carry pink to red flowers in a short, conical raceme.

Flowering is from June to July.

This cliff dweller is evidently closely related to *A. arborescens*. It can be distinguished by its hanging branches and broader, more bluish-green leaves. Unlike *A. mutabilis*, the flowers are always uniformly red, never bicoloured.

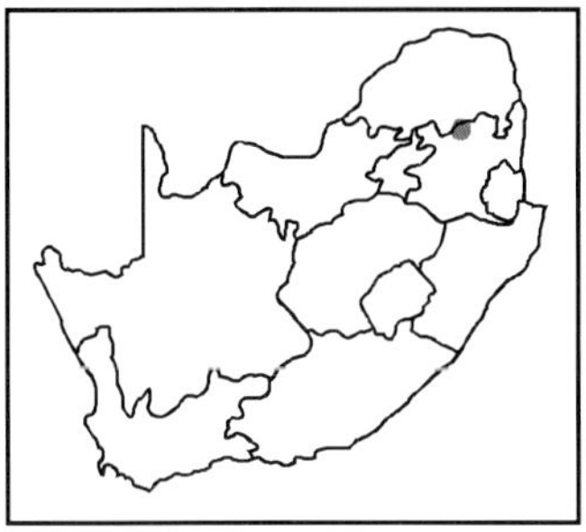

The species occurs in the Lydenburg district of Mpumalanga.

The conservation status of *Aloe hardyi* is insufficiently known.

This species is another recent discovery which is not yet widely cultivated. However, it may have been overlooked in some collections because of its similarity to *A. mutabilis*. The scientific name commemorates David S. Hardy, a well-known plant collector and authority on aloes and other succulent plants.

Peter Raal

Peter Raal

Peter Raal

Aloe mutabilis

Plants are sparingly branched from the base, with short stems that rarely exceed 1 m in length. The stems are erect or more often trailing, particularly when growing against cliffs. The bluish-green leaves are soft in texture and arranged in dense rosettes. The long, slender tips are often shrivelled in mature leaves, probably as a result of frost or drought. Soft, harmless teeth are present along the leaf margins. The inflorescences are usually unbranched or may rarely have two branches. The racemes are up to 1 m high and densely flowered, usually with a striking difference between the colour of the buds (red) and the open flowers (yellow or greenish-yellow). However, the flowers of some forms are uniformly red.

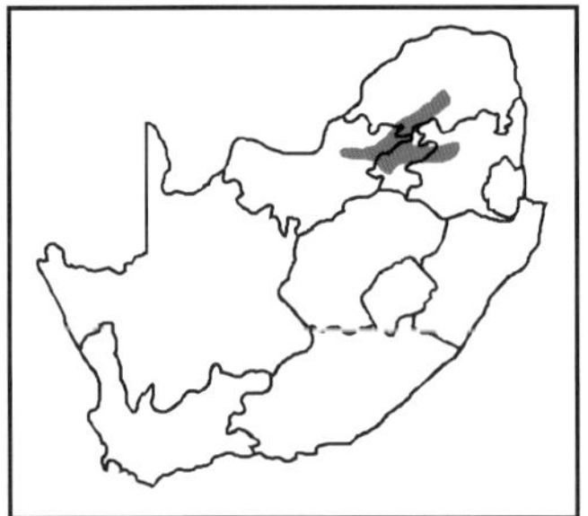

The flowering period stretches from June to August.

Aloe mutabilis is often considered to be merely a Highveld form of *A. arborescens*. It can be distinguished by the absence of a stem and the bicoloured racemes.

The species occurs from Johannesburg in the south, northwards to Chuniespoort near Pietersburg and westwards along the Magaliesberg to Rustenburg.

Aloe mutabilis is not threatened.

The attractive bicoloured racemes and frost resistance make this a popular garden plant on the Highveld. *Mutabilis* means 'changeable' and accurately describes the variability of the species.

Eben van Wyk

Frits van Oudtshoorn

Eben van Wyk

Aloe spicata

The plants are single or multistemmed and up to 2 m high. The spreading recurved leaves have small, firm teeth along the margins and are often reddish-green in colour. Inflorescences are invariably single, but more than one may grow from each rosette. The densely many-flowered racemes have a bottle-brush appearance. The stalkless flowers are small and yellow, with long exserted stamens and dark brown nectar.

Flowering is from July to August.

The species can be distinguished from other multistemmed aloes by virtue of its stalkless flowers and recurved leaves. It can easily be separated from the similar, yet invariably single-stemmed *A. alooides* by the less robust appearance.

Aloe spicata is widely distributed in northern KwaZulu-Natal, southern Mozambique, Swaziland, Mpumalanga and the Northern Province.

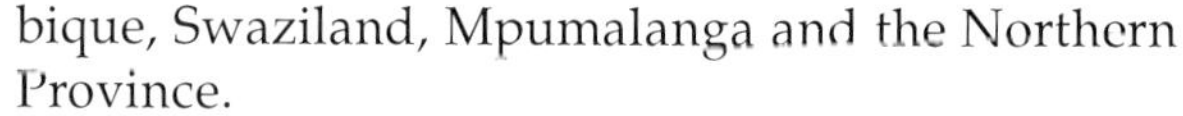

The species is very abundant in its natural habitat and is not threatened.

Aloe spicata was previously known as *A. sessiliflora*. *A. spicata* combines the leaf characteristics of *A. vanbalenii* (recurved, often red in colour) with the densely packed stalkless flowers of *A. alooides* and *A. vryheidensis*. It is an attractive plant in cultivation and will grow well in full sun or dappled shade. The common name is Lebombo aloe or Lebombo-aalwyn. *Spicata* refers to the long and densely flowered, spike-like inflorescences. The national tree number is 30.4.

Ben-Erik van Wyk

Pitta Joffe – NBI

Pitta Joffe – NBI

Aloe succotrina

Mature specimens have branched or unbranched stems of up to 1 m long which are covered with the remains of old, dry leaves. Plants are usually found in groups, solitary specimens being uncommon. The stiff, erect leaves are a dull greyish-green colour and have striking contrasting narrow white borders and prickles on the margins. The inflorescence is a simple or rarely once branched raceme of up to 1 m high. The flowers are up to 40 mm long and bright red.

Flowering is from July to August.

The erect, dull grey-green leaves with their white borders and teeth, and the fairly large purplish flower bracts are the main characteristics separating *A. succotrina* from superficially similar species. The inflorescences are generally unbranched and longer than those of *A. arborescens*.

Aloe succotrina is a distinctive component of the south-western Cape flora. It is one of only a few fynbos aloes and is restricted to mountain slopes, from the Cape Peninsula to Cape Hangklip and Hermanus.

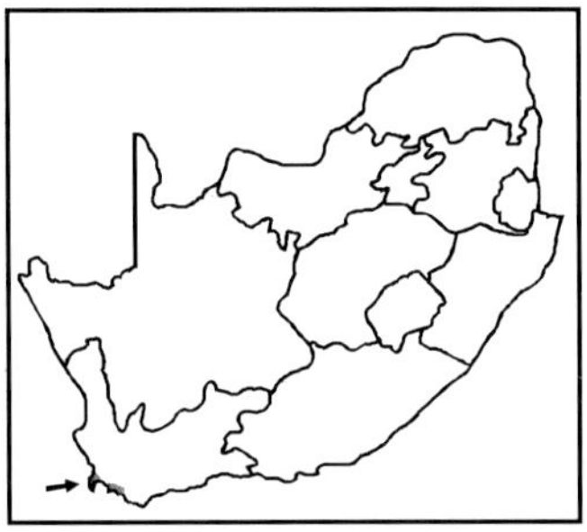

Aloe succotrina is not threatened.

With its dark grey-green leaves and bright red inflorescences, this species is a striking and distinctive aloe. It does not do well in cultivation in the summer rainfall areas of Southern Africa. The species has a confused botanical history: for many years it was thought to have been first collected on the Indian Ocean island of Socotra, hence the rather inappropriate scientific name. There is no doubt today that *A. succotrina* grows wild only in the south-western Cape.

GW Reynolds – NBI

Gideon Smith

Ernst van Jaarsveld – NBI

Aloe vanbalenii

Plants are stemless or will in time form short, robust creeping stems with numerous rosettes. The leaves are strongly recurved and snake-like with the tips touching the ground. If growing in full sun, the deeply channelled leaves are bright red and have marginal teeth of the same colour. In shade, the leaves are bright green. The inflorescences branch from about the middle into two or three narrow, cone-shaped racemes. The flowers are tubular and vary considerably in colour. They are usually different shades of yellow and orange, but occasionally dull red.

The species flowers from June to August.

This distinctive aloe is easily recognised by the strongly recurved and deeply channelled leaves that lie close to ground level. Given the large rosettes, the virtual absence of stems is quite striking.

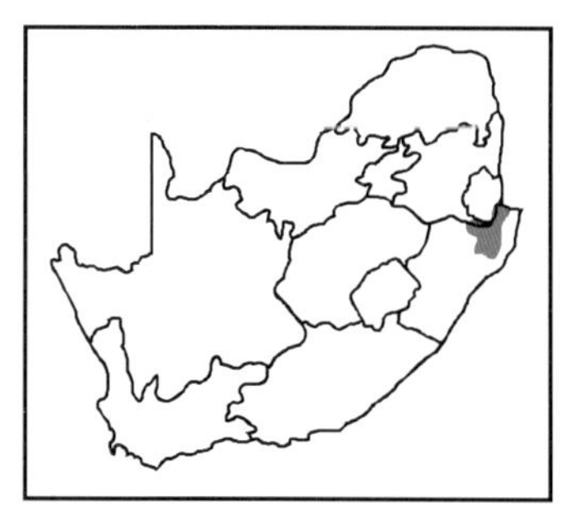

The species occurs in northern KwaZulu-Natal and in the extreme south-eastern part of Mpumalanga.

Aloe vanbalenii is not threatened.

This is one of the most attractive aloes for the garden. It is very easy to cultivate and flowers freely, producing a striking display of narrow, cone-shaped racemes that livens up the often drab winter Highveld garden. The scientific name honours J.C. van Balen, a former Director of Parks of Johannesburg, who first collected the species.

Gerhard Steyn

Leo Thamm

Gerhard Steyn

Rambling aloes

Group 4

This is one of the most distinctive groups of aloes recognised in this book. The species treated here as ramblers are all characterised by having fairly **thin and slender, almost wiry stems** that at length often require at least some support, for example from surrounding trees or shrubs, to remain erect. If grown in exposed positions away from other plants, the stems are usually erectly spreading. The species included here invariably have only **slightly fleshy leaves** which are **widely spaced** on the thin stems. The **sheathing portions** of the **leaves** are usually distinctly **striped**. The **racemes** are fairly **short** and bear comparatively **few,** small to large **flowers**, but the numerous racemes produced by a mature plant compensate for the lack in number of flowers. In their natural habitat the rambling aloes will often develop into large masses of intertwined stems that resemble huge fiery balls when in flower. The group represents a natural assemblage of closely related species.

SPECIES

A. ciliaris
A. commixta
A. gracilis
A. striatula
A. tenuior

Ben-Erik van Wyk

Aloe striatula

Aloe ciliaris

Plants form large masses of semi-woody stems that are supported by surrounding bushes and trees. The stems can reach a length of up to 6 m and are leafless for most of their length. The leaves are mostly borne on the terminal portions of the stems, although in cultivation this could vary depending on the watering regime. The leaves are dark green and well-spaced, and have distinct, white marginal hair-like thorns. These hair-like structures are particularly large on the margins of the leaf sheaths, where they clasp the stems. The inflorescence is usually an unbranched raceme bearing large, yellow-tipped, bright red flowers.

Flowering occurs throughout the year.

The prominent hairy fringes of the sheathing bases of the leaves of *A. ciliaris* make it one of the most distinctive rambling aloes. The large, bright red flowers borne in short, subdense racemes further serve to characterise the species.

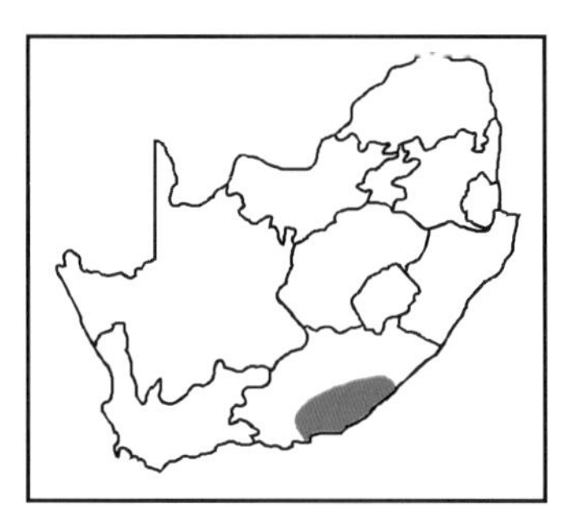

Aloe ciliaris occurs in well-wooded ravines and among bushes in thicket vegetation in the Eastern Cape Province.

The species is not threatened.

Following recent research at the Royal Botanic Gardens in Kew, England, *A. tidmarshii* was reduced to a variety of *A. ciliaris*. The var. *tidmarshii* is very similar to *A. ciliaris* but is less robust in all respects. An additional variety, the var. *redacta*, which is intermediate between the var. *ciliaris* and var. *tidmarshii*, was described at the same time. *Aloe ciliaris* is very easy to cultivate and is widely grown in Southern Africa. If given space to spread, the species will form large, profusely flowering stands. *Ciliaris* refers to the hairy fringes of the leaf sheaths.

Ben-Erik van Wyk

Ben-Erik van Wyk

Ben-Erik van Wyk

Aloe commixta

Plants grow as large shrubs with numerous branched, creeping stems of up to 1 m long. The leaves are erectly spreading and slightly fleshy. The internodes between the leaves are distinctly green-striped. The inflorescences are up to 300 mm high and rather stout for a rambling species. The uppermost floral buds are reddish and are borne erectly spreading, while the open flowers are yellowish-orange to orange and point downwards.

Aloe commixta flowers in August and September.

Although more robust than in most rambling species, the stems of *A. commixta* usually tend to become sprawling. The rather erectly borne leaves of the species do not become as recurved as in *A. striatula*, the species to which it is most closely related. Its inflorescences are shorter, less robust and the racemes are not as distinctly cone-shaped as in *A. striatula*. There is also a clearer dispositional distinction between the buds (upright) and open flowers (hanging) of *Aloe commixta*.

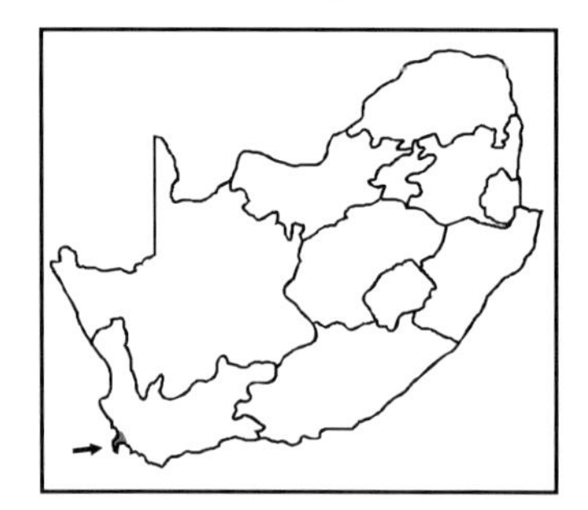

Aloe commixta has a restricted distribution on the Cape Peninsula.

The species is not threatened.

Aloe commixta does not grow well outside of its natural habitat. It prefers the acidic, sandy soils and the winter rainfall regime of the south-western Cape. *Commixta* probably refers to the typically 'intermingled' stems of the plant.

Ben-Erik van Wyk

Ben-Erik van Wyk

Ben-Erik van Wyk

Aloe gracilis

Plants form large shrubby stands of semi-erect stems. The plants branch from near ground level and may reach a length of about 2 m. The internodes on the stems are up to 15 mm long and have faint green striations. The leaves are dull green and narrow, and are borne rather erectly spreading. The leaf margins are armed with white teeth of about 1 mm long. Inflorescences are once or twice branched and up to 300 mm high, but usually only the upper third carries laxly dispersed flowers. The comparatively large, deep red flowers are borne pendulously at maturity.

Flowering is from May to July.

Aloe gracilis appears to be most closely related to *A. striatula*, from which it differs in having large red flowers borne in sparse racemes. The leaf internodes of *A. gracilis* are also not as distinctly striped as those of *A. striatula*. *Aloe gracilis* differs from *A. ciliaris* in that it lacks the marginal hair-like fringe around the stem-clasping part of the leaves and has erect, not recurved leaves.

Typically, *A. gracilis* is confined to thicket vegetation in the Port Elizabeth and Uitenhage districts in the Eastern Cape Province. *Aloe gracilis* var. *decumbens*, discussed below, occurs in the Langeberg in the southern Cape.

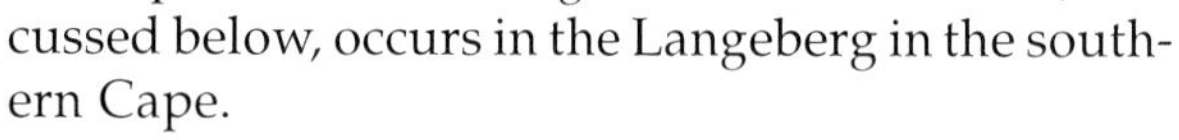

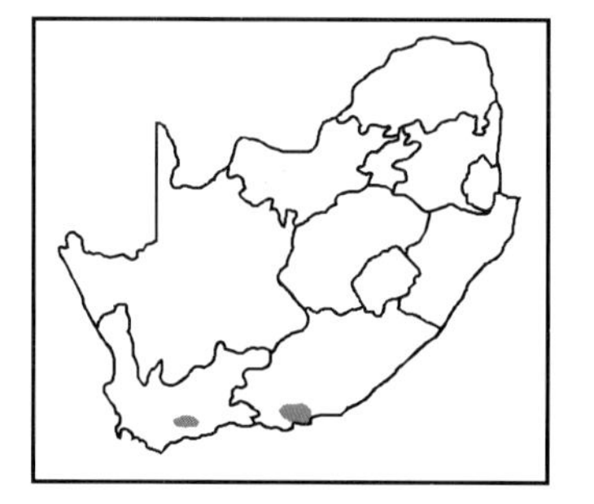

Aloe gracilis is not threatened, but the var. *decumbens* is becoming increasingly rare in its natural distribution range.

A single variety of *A. gracilis*, the var. *decumbens*, has been described. The main differences between the variety and typical *A. gracilis* are that the former is a less robust plant in all respects, with a more sprawling habit. It is unlikely that this variety will be upheld after in-depth study. *Aloe gracilis* is easier to cultivate than *A. commixta*, which has a distinct preference for a winter rainfall regime. The scientific name was well chosen, because *gracilis* means thin and slender, obviously referring to the stems of the plant.

Ben-Erik van Wyk

Ben-Erik van Wyk

Ernst van Jaarsveld – NBI

Aloe striatula

Plants are robust and form large, untidy shrubs of up to 2 m high and often several metres wide. The stems are more or less 20 mm in diameter, with the portions between the widely spaced leaves distinctly marked with green longitudinal lines. The leaves are usually deeply channelled, shiny dark green in colour and weakly to strongly recurved. A substantial part of the leaves of some forms will die back into a thin, black shrivelled portion. This phenomenon is usually associated with low temperatures, but in *A. striatula* it occurs throughout the year. The inflorescences are unbranched racemes of up to 400 mm high, although this length is hardly ever reached in cultivation. The flowers are fairly large and pendulous, and tend to be tightly packed around the peduncle.

Flowering is from November to January.

Aloe striatula is distinguished from other rambling aloes by its recurved, shiny, dark green leaves. It is also the most robust rambling aloe and has slightly curved flowers that are borne in a relatively dense raceme.

The species occurs from west of Graaff-Reinet in the Karoo to Queenstown and Lady Grey in the eastern and north-eastern parts of the Eastern Cape Province, and the mountains of southern Lesotho.

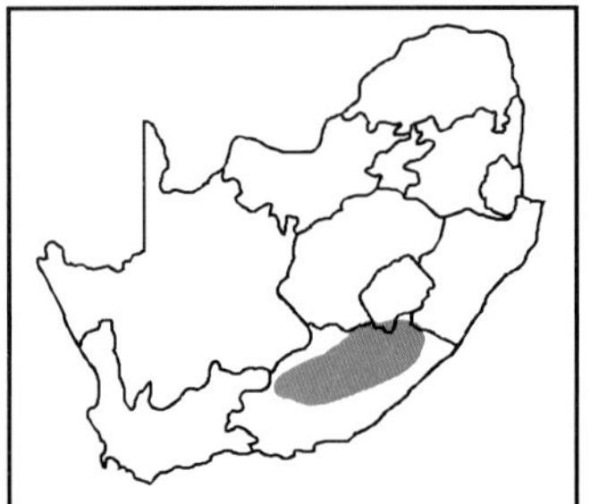

Aloe striatula is not threatened.

A single variety of *A. striatula*, the var. *caesia*, is recognised. It differs from the typical variety in having more spreading, closely packed, greyish-green leaves and shorter, straighter flowers. *Aloe striatula* is a useful garden plant that will in time form large multistemmed bushes. However, many of the cultivated forms of the species do not flower well. Its horticultural potential lies in its foliage and the fact that it can tolerate extremely low temperatures; it will grow happily in areas where few other aloes will survive. The Latin *striatula* refers to the thin, green, parallel lines on the leaf sheaths.

Ben-Erik van Wyk

Ben-Erik van Wyk

Ben-Erik van Wyk

Geoff Nichols

Aloe tenuior

Plants form small to medium-sized untidy bushes. Stems are up to 3 m high with the leaves crowded in more or less laxly leaved, terminal rosettes. The stems arise from a very large rootstock that becomes somewhat woody near ground level. The leaves are greyish-green and unspotted, with small marginal teeth. The racemes are usually unbranched and vary from laxly to densely flowered. Flower colour ranges from yellow to red.

Although peaking from early winter (May) right through to late winter (August), flowering does not appear to be restricted to any particular time of the year.

In the group of rambling aloes, *A. tenuior* is characterised by its large rootstock, blue-green leaves and fairly long racemes with small red or yellow flowers. It also lacks the fringed leaf sheaths of *A. ciliaris* and its varieties.

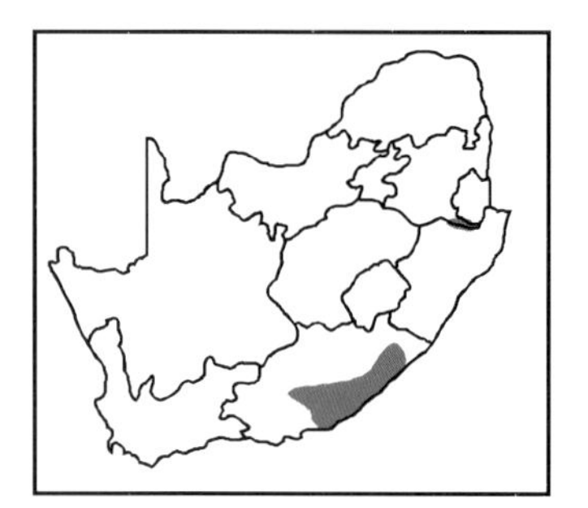

The species occurs in a broad belt from Cookhouse and Somerset East in the Eastern Cape Province northwards to Tsolo in southern KwaZulu-Natal.

Aloe tenuior is not threatened.

This aloe is a very useful landscape plant that will in time form large, bushy masses of slender stems. In cultivation it is free-flowering and produces large numbers of yellow or red flowers. Three varieties, var. *decidua*, var. *densiflora* and var. *rubriflora*, have been recognised in the past. These are all best treated as extreme variations of a very variable species. *Tenuior* means very thin, obviously referring to the slender branches.

Ben-Erik van Wyk

Geoff Nichols

Ben-Erik van Wyk

Creeping aloes

Goup 5

This distinctive group of aloes consists of seven species that are distinguished chiefly by the **creeping habit of the stems**, or if stems are absent or indistinct, by the **rosettes that are tilted to one side**. Leaf shape differs markedly amongst the species, but there is a tendency towards a **dull greenish leaf colour**, with the margins carrying **harmless, white teeth**. All the species of creeping aloe **flower in the dry summer months of the Northern, Western and Eastern Cape Provinces**. The **inflorescences** of all the species included in this group are more or less **head-shaped** and **many-flowered**. It is noteworthy that five of the seven creeping aloes are threatened in one way or another. Two species, *A. dabenorisana* and *A. meyeri*, treated here as creeping aloes, were described as recently as the early 1980s, after the publication in 1950 (and its updates in 1969 and 1974) of *The aloes of South Africa* by Gilbert W. Reynolds.

SPECIES

A. arenicola
A. comptonii
A. dabenorisana
A. distans
A. meyeri
A. mitriformis
A. pearsonii

Ben-Erik van Wyk

Aloe distans

Aloe arenicola

The plant forms dense, medium-sized shrubs consisting of numerous creeping stems. The leafy portions of the stems are erect, whether supported by surrounding bushes or not. The fairly narrow, upturned leaves are up to 200 mm long and quite densely arranged in terminal rosettes. As with most creeping aloes, the leaves are bluish-green and have numerous whitish spots on both surfaces. The leaf margins have whitish edges and are armed with very small teeth. The inflorescences are usually simple, but are sometimes once or twice branched. The pale red flowers are borne in tightly packed head-shaped racemes.

Flowering occurs from July (in the northern part of its distribution) to December (the southern part of its distribution).

Aloe arenicola can be easily distinguished from other creeping aloes by its narrower, bluish-green leaves that have numerous white spots on both surfaces. Based on the age of specimens, two distinct forms of the species are found; that is young and mature plants differ markedly especially in terms of leaf shape. These differences are discussed below.

Aloe arenicola is confined to the arid coastal strip of sandveld on the west coast of South Africa, stretching from Lambert's Bay in the south to the mouth of the Orange River in the north (Namibian border). This distribution cuts across the border dividing the Western and Northern Cape Provinces.

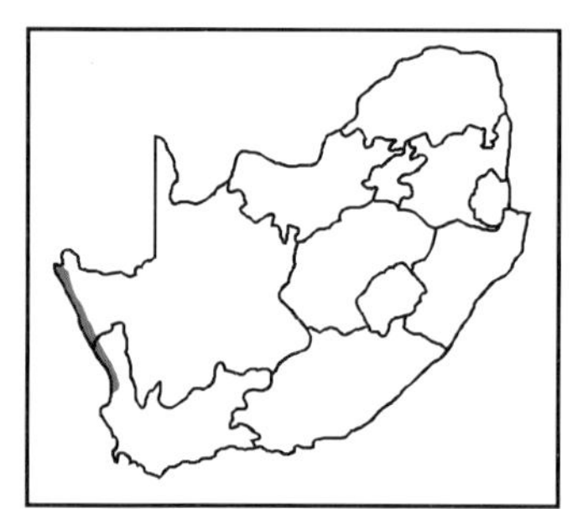

The species is vulnerable, due to mining activities and overgrazing.

This is a very distinctive species, especially in the juvenile stage. The young plants have thin, creeping stems with small, very succulent, almost triangular leaves. The internodes between the leaves are much longer and more obvious in young plants. In contrast to *A. distans* and *A. mitriformis*, only the terminal 200 mm of the stems of *A. arenicola* bear leaves. The scientific name of the species refers to its occurrence in sandy habits (*arenicola* means 'inhabitant of sandy places').

Nick Klapwijk – NBI

Nick Klapwijk – NBI

Ernst van Jaarsveld – NBI

Aloe comptonii

Plants are either stemless or could develop creeping stems of up to 1 m long. Especially in old specimens, the stems tend to be horizontal rather than erect. The leaves are dull bluish-green and up to 300 mm long. They are narrow and triangular in shape. The leaf margins are armed with short, firm, white teeth. The inflorescence is once or twice branched, consisting of densely flowered racemes. Individual racemes are fairly short and head-shaped. Flower colour varies from dull to bright red.

Flowering is from August to January.

Aloe comptonii is the most robust of all the creeping aloes. It is characterised by its larger leaves and large, head-shaped racemes. In contrast to *A. arenicola*, the leaf prickles are not carried on a white, horny margin. The flowers are fairly thin and tightly packed in a head-shaped raceme. The individual flower stalks are rather thin and point upwards, whereas the flowers are pendulous.

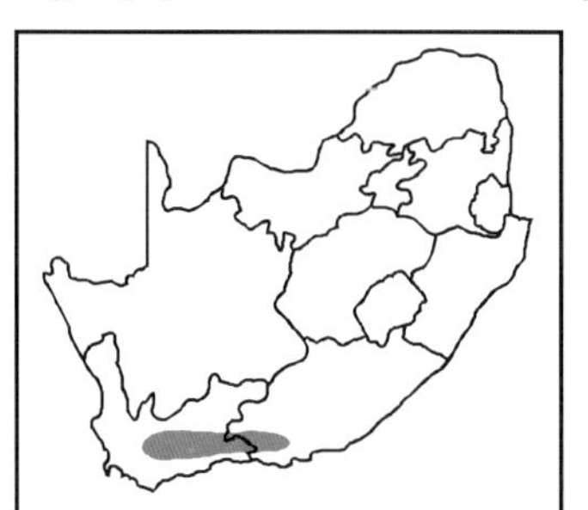

The species has a wide distribution in the karroid regions of the Western and Eastern Cape Provinces. Its distribution stretches from Montagu in the west to near Uitenhage in the east.

Aloe comptonii is not threatened.

Although the species might at first glance appear fairly distinctive, it merges with the less robust and more distinctly creeping *A. mitriformis* in the western extremes of its distribution range. A critical study might well prove these two species to be extreme variations of a single, variable species. *Aloe comptonii* is very easy to cultivate in areas that are close to its natural distribution range. However, it will not survive the extremely low winter temperatures that are often encountered on the Highveld of the northern provinces of South Africa. The species was named after Prof. R.H. Compton, second director of the National Botanical Gardens of South Africa.

Ben-Erik van Wyk

Ben-Erik van Wyk

Marinda Koekemoer – NBI

Aloe dabenorisana

Plants characteristically hang down from the vertical cliff faces against which they grow. The stems often branch to form multiheaded clusters of up to 1 m in diameter. The dull green leaves are strongly recurved, that is bent backwards towards the stem, and usually have a reddish tinge. The leaf margins are armed with small white triangular teeth. The inflorescences are typically branched, consisting of two to four cone-shaped racemes. They grow down from the pendulous rosettes, with the terminal portions that carry the flowers turned upwards. The flowers are yellowish with a red line along the middle of each individual segment.

Aloe dabenorisana flowers from August to November.

At first glance the species is very similar to *A. meyeri*. However, it differs from the latter by being much larger and stemless. In addition, the leaves of *A. dabenorisana* are recurved, deeply channelled and the inflorescences are more consistently branched. In the case of *A. meyeri*, the leaves are erectly spreading and not as deeply channelled, and the inflorescences are only very rarely branched. A cha-racteristic which *A. dabenorisana* shares with *A. pearsonii*, also a creeping aloe, is that the leaves of juvenile plants are arranged in four to five vertical rows.

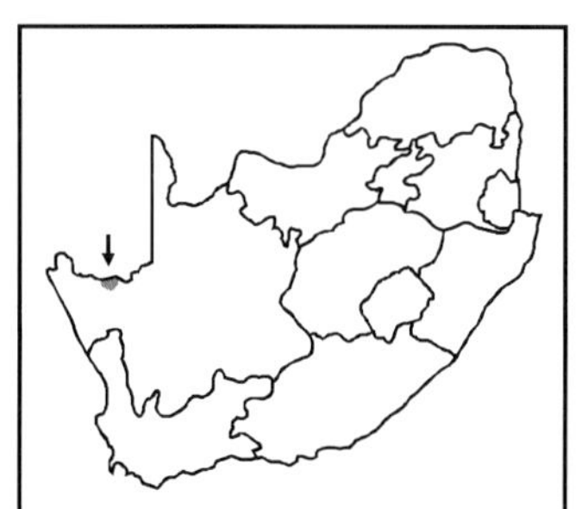

Aloe dabenorisana has a limited distribution range in northern Bushmanland where it is restricted to quartzitic rock crevices in the Dabenoris Mountains.

Aloe dabenorisana is rare due to illegal collection and a small world population.

This very distinctive species is little known in cultivation and has not yet been widely introduced into horticulture. It grows in an extremely arid region and it is likely that in cultivation it will pose some difficulties, even to the most dedicated of collectors and growers of aloes. *Aloe dabenorisana* was named after the Dabenoris Mountains where it was first collected.

Ernst van Jaarsveld – NBI

NBI

Aloe distans

Plants have a distinctly creeping growth habit, forming dense groups of many sprawling stems and individual rosettes. Stems may reach a length of 3 m and give rise to numerous offshoots. The slightly fleshy, bluish-green leaves are up to 150 mm long and more or less triangular in shape. Both surfaces have whitish spots, the lower surface more so than the upper. The leaf margins are armed with firm, yellowish teeth. The inflorescence is branched, carrying up to four short, tightly packed racemes. The densely flowered racemes are distinctly head-shaped. Flower colour varies from a dull orange-red to bright red.

This is a summer flowering species that produces its inflorescences in December, at the height of the western Cape dry season.

Aloe distans has few distinguishing characteristics, apart from having less robust stems and leaves than *A. mitriformis*. It is to some extent characterised by its very long, creeping stems covered with the remains of old, dry leaves. The flowers are similar to those of *A. mitriformis*.

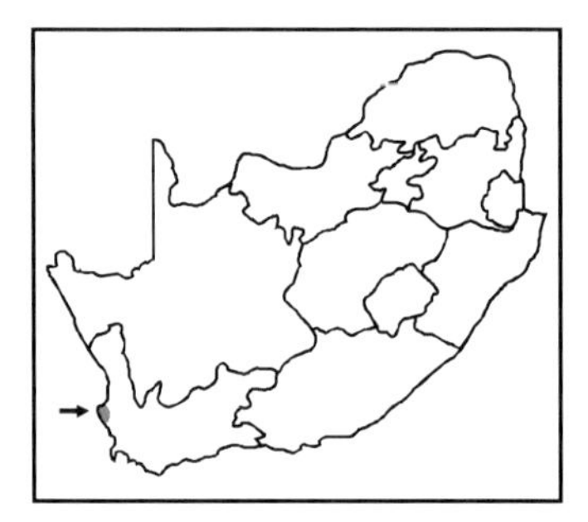

Aloe distans has a fairly restricted distribution along the west coast of South Africa, from Danger Point northwards to St Helena Bay.

The species is regarded as rare.

Aloe distans is not widely cultivated and appears to have very specific horticultural requirements since it does not do well outside of its natural habitat. The name *distans* means 'standing apart' or 'far removed', which is more true of the geographical distribution than of the general appearance of this aloe. Like *A. comptonii*, this species might well be little more than a variation of *A. mitriformis*, but in this case with a preference for a coastal climate.

Ben-Erik van Wyk

Ben-Erik van Wyk

Ben-Erik van Wyk

Aloe meyeri

Like *A. dabenorisana, A. meyeri* also grows hanging down from vertical cliff faces. Stems are up to 1 m long and branch from the base or have offshoots forming along the length of the stem. The bluish-green leaves are erectly spreading and up to 300 mm long. Dry leaves remain on the stems for a short time only. The leaf margins are armed with small, white teeth. The inflorescences are simple, head-shaped racemes that curve upwards from the pendulous rosettes. The flowers are reddish-orange and have green tips.

Aloe meyeri flowers from December to February.

Aloe meyeri is a much smaller plant than *A. mitriformis,* from which it also differs in the hanging rosettes and upcurved racemes. Geographically *A. meyeri* occurs in the vicinity of the Rosyntjieberg in the Northern Cape Province, whereas *A. mitriformis* is a south-western Cape species. (Differences between *A. meyeri* and *A. dabenorisana,* which also occurs on vertical cliff faces, are discussed under the latter.)

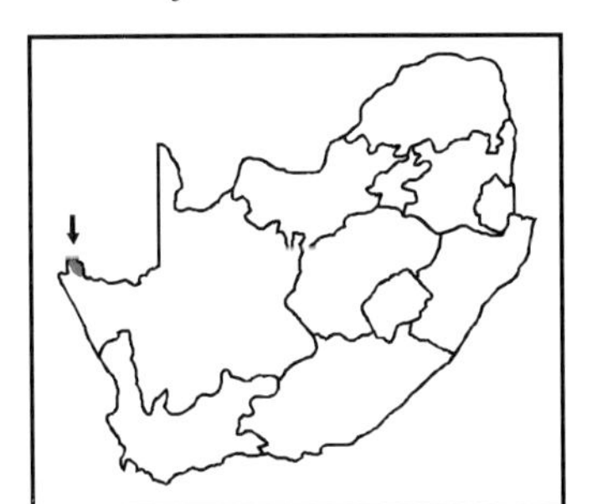

The species has a restricted distribution range in the Northern Cape Province where it is confined to the Rosyntjieberg in the Richtersveld. *Aloe meyeri* also occurs in Namibia where this mountain range is dissected by the international border.

Due to its very localised distribution *A. meyeri* is regarded as rare.

As in the case of *A. dabenorisana,* the other recently described creeping aloe, *A. meyeri,* has not yet been widely introduced into cultivation. These species are two of the few pendulous aloes that occur in Southern Africa. *Aloe meyeri* was named in honour of Rev. G. Meyer of Steinkopf who first collected specimens of this species.

Graham Williamson

Ernst van Jaarsveld – NBI

Pitta Joffe – NBI

Aloe mitriformis

Plants develop long, creeping stems that give rise to side-shoots along their lengths. These sprawling stems are up to 2 m long and typically lie along the ground with just the terminal leaf-bearing portion being erect. The thick, fleshy leaves are bluish-green and erectly spreading to incurved. The leaf margins are armed with small, white teeth that turn dark brown to yellow on old leaves. The inflorescences are branched, and consist of up to five densely flowered racemes. The shape of the racemes varies from cone-shaped to head-shaped and rounded. The flowers are comparatively slender, and dull to bright red in colour.

Aloe mitriformis flowers from December to February.

The species can be regarded as the most typical of the creeping aloes. *Aloe mitriformis* is very variable and it is difficult to single out specific characteristics that distinguish it from other creeping aloes. The slightly incurved leaves and variably coloured marginal spines appear to be consistent features of the species. The stems of *A. mitriformis* are fairly thin and not strong enough to support the large rosettes in an erect position.

Aloe mitriformis is widely dispersed in the south-western Cape, from near Genadendal in the south to the Bokkeveld Mountains near Nieuwoudtville in the north. It typically occurs in flat, rocky places, although it is not uncommon on vertical cliffs.

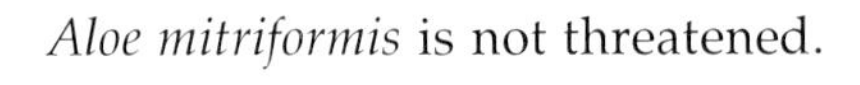

Aloe mitriformis is not threatened.

The name of the species is derived from the resemblance of the rosette to a Mitre or Bishop's cap, especially in times of drought. The common names of *Aloe mitriformis* are Mitre aloe and Kransaalwyn. The latter name is indiscriminately applied to a number of unrelated species, such as *A. arborescens*. If *A. distans* and *A. comptonii* are included with *A. mitriformis*, then the latter name will take precedence (see the former two species for discussions).

Craig Hilton-Taylor

Ernst van Jaarsveld – NBI

JLC Marais

Aloe pearsonii

Plants grow as large, much-branched shrubs of up to 2 m in diameter. Branching takes place from the base of the stems or higher up. The stems are erect and carry leaves for most of their length. The dull bluish-green leaves are almost triangular in shape and often take on a red colour, especially in times of drought. They are neatly arranged in vertical rows and curve backwards, giving the rosettes a distinctive appearance. The inflorescence is often branched low down into two or three head-shaped racemes. The flowers are red to orange-red or yellow.

Aloe pearsonii flowers from December to January.

This is probably the least typical of the creeping aloes. In contrast to most of the other species, the stems are usually erect and only rarely become creeping. The leaves, which are the smallest in the group, cover the stems for most of their lengths. At localities where *A. pearsonii* occurs abundantly, the reddish-green leaves will impart a reddish colour to the surrounding hillsides.

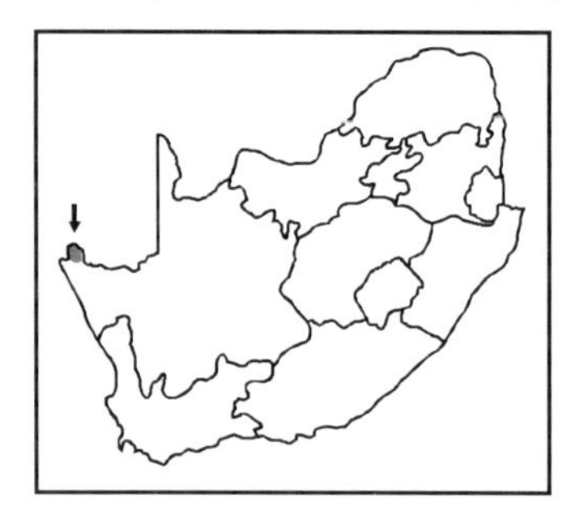

Aloe pearsonii is vulnerable due to mining activities, overgrazing and illegal collecting.

Aloe pearsonii grows naturally in one of the most arid parts of South Africa. It is not easy in horticulture and watering should be done carefully if it is to be kept successfully in cultivation. The species was named after Prof. Harold H.W. Pearson, the first director of the then National Botanical Gardens of South Africa.

Graham Williamson

Edgar Swart

Pitta Joffe – NBI

Stemless aloes

Group 6

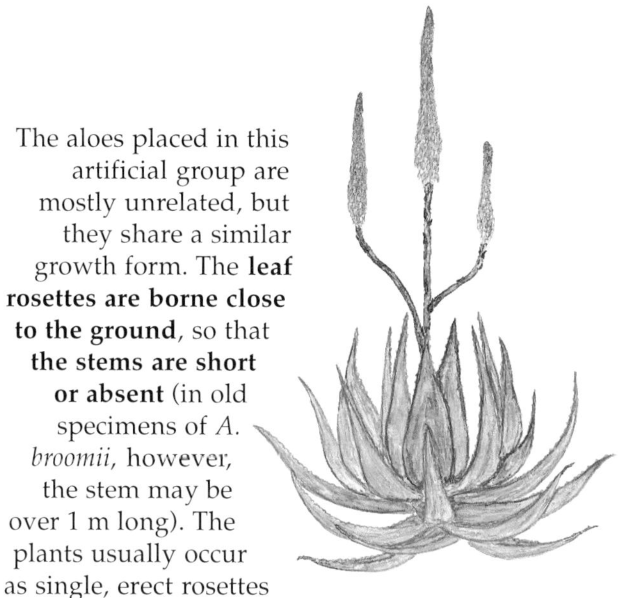

The aloes placed in this artificial group are mostly unrelated, but they share a similar growth form. The **leaf rosettes are borne close to the ground**, so that **the stems are short or absent** (in old specimens of *A. broomii*, however, the stem may be over 1 m long). The plants usually occur as single, erect rosettes but in *A. claviflora*, *A. falcata*, *A. haemanthifolia* (and to some extent also *A. pratensis*), the rosettes are slanted to horizontal and they form dense clusters in mature specimens. The inflorescences and flowers vary greatly but are relatively large in relation to the small size of the plants. They are also very showy, so that some of the most distinctive and spectacular aloes are grouped here.

SPECIES

A. aculeata
A. broomii
A. buhrii
A. chabaudii
A. chlorantha
A. cryptopoda
A. gerstneri
A. glauca
A. globuligemma
A. haemanthifolia
A. hereroensis
A. krapohliana
A. lutescens
A. melanacantha
A. peglerae
A. petricola
A. polyphylla
A. pratensis
A. reitzii
A. reynoldsii
A. striata
A. suprafoliata
A. thorncroftii

Beth Peterson

Aloe peglerae

Aloe aculeata

Plants form single rosettes of up to 1 m high and wide. The leaves are long and relatively broad, up to 120 mm wide at the base, with the upper parts distinctly curved inwards to give the rosette a rounded appearance. Reddish-brown triangular teeth occur along the leaf margins. A distinctive feature is the numerous thorns on the upper and lower leaf surfaces. Each thorn arises from a thick, tuberculate base and in some forms of the species the tuberculate part of the thorn is paler than the surrounding leaf, giving the leaf surfaces a characteristically dotted appearance. The inflorescences are single in young plants but may have three or four branches in old plants. Each raceme is long and narrow, and gradually tapers towards the tip. The flowers are tubular, up to 40 mm long and vary from yellow to reddish-orange. Some forms are uniformly yellow or orange, while others are bicoloured, with the flowers orange in the bud stage and yellow when open.

Flowering occurs from May to July.

Aloe aculeata differs from similar species by the conspicuously tuberculate spines on the leaves. This characteristic, in combination with the relatively large, single rosettes and long, narrow, erect racemes, makes the species rather easy to identify.

This aloe occurs in several parts of the Northern Province and also in the extreme northern part of Mpumalanga. Northwards, the distribution extends into Zimbabwe. It is invariably found in rocky areas, usually in grassland or open bushveld.

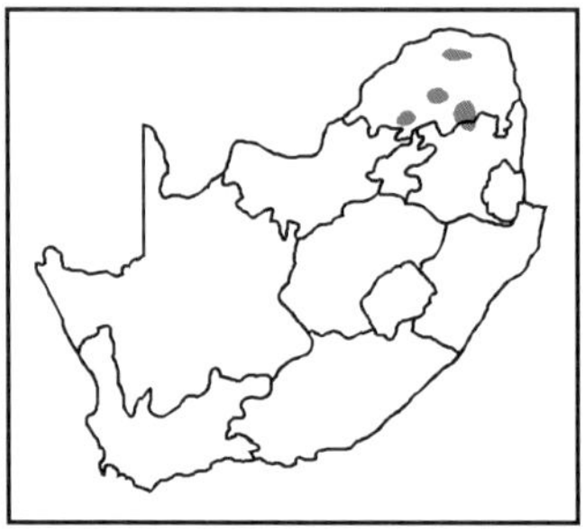

The species is not threatened.

Aloe aculeata is well known to many South Africans because it was depicted on the now discontinued nickle 10 cent coin. This attractive aloe deserves to be planted in gardens more often. The scientific name *aculeata*, which means 'prickly', accurately describes the leaf surfaces. The common names Ngopanie and Sekope have been recorded.

Eben van Wyk

Eben van Wyk

Pitta Joffe – NBI

Aloe broomii

Plants are usually solitary or may rarely be branched into two or three heads. Stems are usually short and invisible, but erect stems of nearly 1 m high may develop in old specimens. The leaves are relatively short and broad, yellowish-green in colour and densely crowded in the rosette. Each leaf is about 300 mm long and ends in a narrow, dry, spine-tipped upper portion. Small, sharp, triangular teeth occur along the brown horn-like margins. The leaf surfaces are yellowish-green with indistinct longitudinal lines, but there are no spots or thorns. The inflorescences are usually unbranched, with one or two arising from each rosette. The racemes are exceptionally long and narrow, up to 1 m long and only about 70 mm in diameter. The flowers are short and broad, up to 25 mm long and pale yellow in colour. A peculiar feature is that the open flowers are completely hidden by large bracts, so that only the stamens are visible (the exposed portion becomes dark orange). However, in the var. *tarkaensis* the bracts are much shorter, so that the flowers are exposed.

The usual flowering time is from late August to early October. In the var. *tarkaensis*, however, flowering occurs in February and March.

This distinctive aloe can easily be recognised by the compact, densely leafy rosettes and particularly by the long, slender, snake-like racemes. (Differences between *A. broomii* and the highly localised *A. chlorantha* are given under the latter.)

Aloe broomii occurs on rocky slopes over a large area of the central interior of South Africa. The distribution area includes parts of the Western, Eastern and Northern Cape Provinces and extends into the southern Free State and western Lesotho.

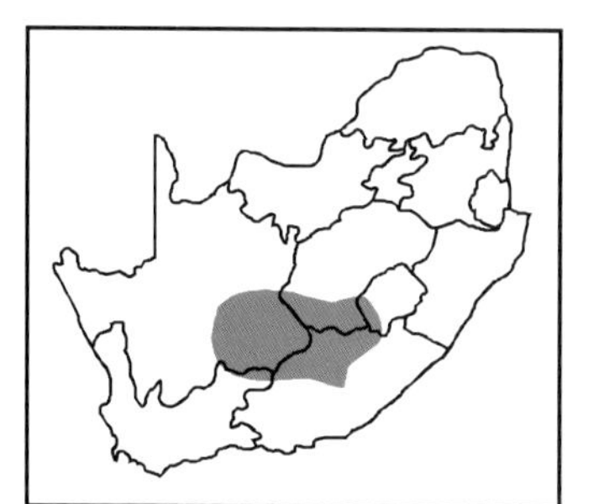

The species is exceptionally common and is not threatened.

Two distinct varieties of *A. broomii* can be distinguished: (1) the widespread typical variety, easily recognised by the large bracts which completely obscure the flowers; (2) the var. *tarkaensis*, characterised by small bracts and therefore conspicuous flowers. The latter is restricted to the extreme south-eastern parts of the distribution area. The species is named after the well-known anthropologist Robert Broom, who first collected it. The common names are Bergaalwyn and Slangaalwyn.

Bill van der Riet

Brian Kemble

Brian Kemble

Joggie van Staden

Aloe buhrii

Plants are stemless. Mature specimens will divide to form large, multiheaded, low-growing groups. The leaves are erect and oblong-triangular. Leaf colour varies from yellowish-green to glaucous-green and the leaves are distinctly marked with H-shaped spots. The reddish leaf margins vary from smooth to minutely toothed. The inflorescence is sparse, with up to fifteen more or less head-shaped racemes. The flowers are orange-red or rarely yellow, and hang down at maturity.

Aloe buhrii flowers from August to October.

The species is similar to *A. striata* but can easily be recognised by its narrower, more upright, yellowish-green leaves. The leaves are also thicker and more succulent. The inflorescences of *A. buhrii* are divided into several small racemes.

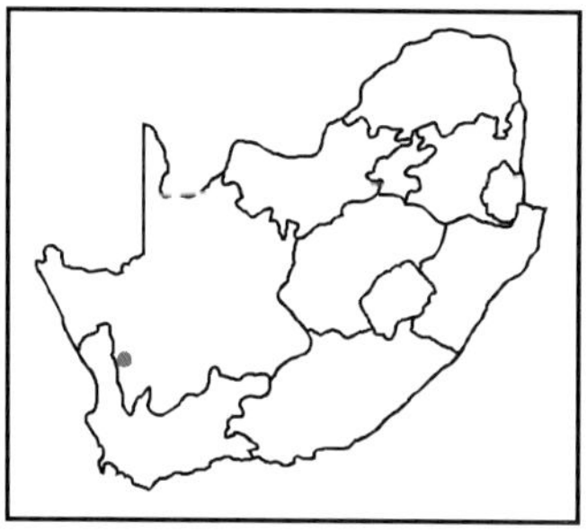

Aloe buhrii occurs in a small, low rainfall area near Calvinia in the Northern Cape Province.

Aloe buhrii is regarded as rare due to its small world population.

The species is not very common in succulent plant collections and it is relatively difficult to cultivate successfully. It prefers winter rainfall and is prone to black, rusty leaf spots in summer rainfall regions. The scientific name commemorates Elias Buhr, a farmer from the Nieuwoudtville district who first collected the species.

Ben-Erik van Wyk

Ben-Erik van Wyk

Ben-Erik van Wyk

Aloe chabaudii

Plants are stemless and produce numerous suckers. The leaves are greyish-green and are borne in erect rosettes. In the case of young plants the leaves could be white-spotted on both surfaces. In mature specimens the leaves are usually without spots but have faint lines on both surfaces. The leaf margins carry well-spaced, short, firm white teeth. The species has much-branched inflorescences, with the flowers in spreading or, more commonly, head-shaped racemes. The flowers are usually red or pinkish and somewhat resemble those of the spotted aloes.

Aloe chabaudii flowers in June and July.

The leaves of *A. chabaudii* tend to be more greyish in colour and are less recurved than in *A. globuligemma*, which is probably the only species with which it could be confused when not in flower. It is generally also a smaller plant than *A. globuligemma*. The inflorescences of *A. chabaudii* are usually spreading, with head-shaped or cone-shaped clusters of reddish flowers. In the case of *A. globuligemma*, the side-branches of the inflorescences are borne horizontally. The deep pinkish flowers of the latter species are globular and club-shaped, and are borne more or less vertically.

Aloe chabaudii is widely dispersed in Tanzania, Malawi, Zambia, Zimbabwe and Swaziland. In South Africa it occurs in the Northern Province, Mpumalanga and in KwaZulu-Natal. It is a species of the warmer grassland or bushveld regions and occurs in the open or among bushes.

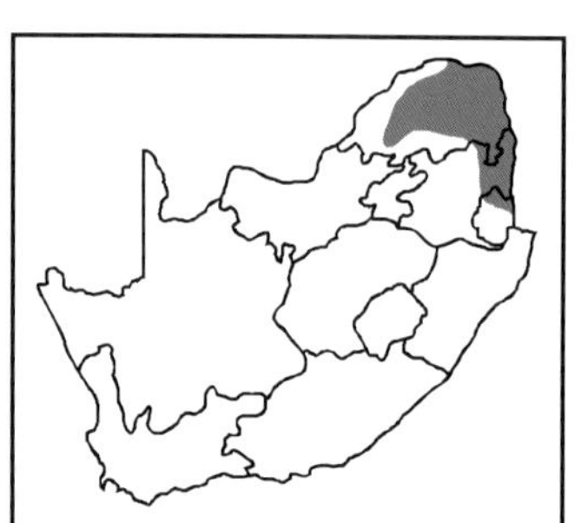

Aloe chabaudii is not threatened.

This is a medium-sized aloe with neat rosettes that will in time form large clusters in a garden. Yellow and orange flowering forms of *A. chabaudii* have been collected in Zimbabwe and Malawi. Two varieties of the species, the var. *mlanjeana* and var. *verekeri*, have been described. The former is restricted to Malawi and is characterised by the smaller rosettes, bright green leaves with white lines on the leaf margins, and inflorescences that are usually shorter than in the typical form. The var. *verekeri* occurs in Zimbabwe and can be distinguished by its more or less head-shaped racemes and olive-green leaves with hooked marginal teeth. Considering the immense variability encountered in *A. chabaudii*, it is debatable whether these varieties should be upheld. The original specimens flowered in the garden of John A. Chabaud, after whom the species was named.

Darryl Plowes

Ben-Erik van Wyk

Pitta Joffe – NBI

Aloe chlorantha

Young plants are solitary but older ones divide into two, three or four rosettes (some plants may have up to ten heads). Stems are usually absent or when present, they grow along the ground and are therefore not readily visible. The leaves are bright green, but may turn various shades of pink and purple when under drought stress. The margins have sharp brown teeth, while the upper and lower surfaces are without any spines or prickles. Some whitish spots are often present. The inflorescences are usually unbranched and over 1 m long. Up to three long, narrow racemes may arise from each rosette. The flowers are very small, only about 10 mm long and yellowish-green in colour. The buds are hidden by the bracts, but the open flowers are visible despite their small size.

Aloe chlorantha flowers in October.

The species is similar to *A. broomii* but may be distinguished by the clustered growth form and the smaller yet more conspicuous flowers, which are not hidden by bracts at maturity. It is also similar to *A. gariepensis* but the latter has much larger flowers (over 20 mm long) and the leaves have distinct longitudinal lines.

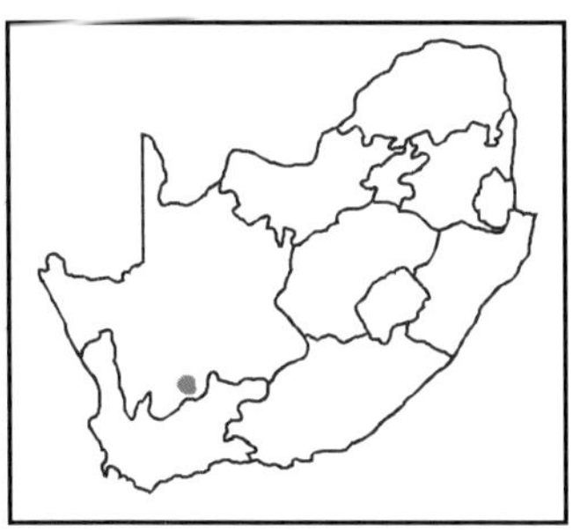

This species occurs on dry rocky northern slopes in a very small area near Fraserburg in the Northern Cape Province.

As a result of the highly localised distribution and susceptibility to insect damage, *A. chlorantha* is considered an endangered species.

The leaves, inflorescences and geographical distribution of *A. chlorantha* are somewhat intermediate between those of *A. broomii* and *A. gariepensis*. Chemical compounds in the leaf sap indeed show a definite link with *A. broomii* and also with several species of speckled aloes (Group 7), the group which includes *A. gariepensis*. *Chlorantha* means 'green flowers', referring to the distinctive flower colour of the species.

Ernst van Jaarsveld – NBI

Ernst van Jaarsveld – NBI

Ernst van Jaarsveld – NBI

Aloe cryptopoda

This aloe grows as a single, more or less stemless, densely leaved rosette. The leaves are narrow and oblong, somewhat greyish-green, without any spines on the surface. There are small, sharp, reddish-brown teeth along the margins. The inflorescences are each branched into three to eight racemes. The racemes are narrow and oblong or broad and conical, and vary from 200 to 400 mm in length. Three main colour forms can be distinguished, namely uniform red, uniform yellow or bicoloured (see below). Each flower is tubular and straight, or somewhat upturned at the mouth, and about 30 to 40 mm long.

The flowering time of the species is variable, but the main period is June and July. At some localities the plants flower in May, or even as early as February or March.

Aloe cryptopoda is a variable species and we here include the two main colour forms of what was previously called *A. wickensii*. The leaf sap chemistry of all these forms is identical, providing further evidence that they are no more than variations of a single variable species. *Aloe cryptopoda* is very closely related to *A. lutescens*, but may be distinguished by the larger, more robust rosettes, and by the broader and more densely flowered racemes. The two species are also geographically separated, with *A. lutescens* occurring further north than *A. cryptopoda*.

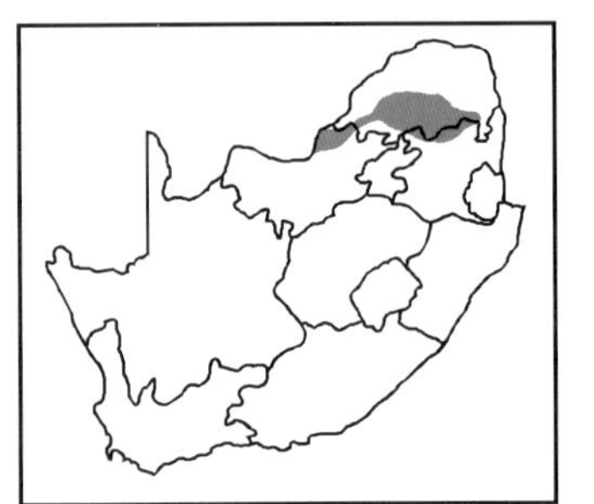

The species is widely distributed in bushveld areas, in flat open places or on rocky slopes. In South Africa it occurs in a broad band across the subtropical part of the country, more or less following the southern boundary of the Northern Province.

Aloe cryptopoda is not threatened.

This species is a popular garden plant and provides a spectacular display when in full flower. For convenience, the three main colour forms should perhaps be formally recognised as varieties. These are: (1) the robust form with oblong, red racemes (the buds and flowers are all red); (2) the smaller form with broad, cone-shaped, bicoloured racemes (the buds are red but the open flowers turn yellow); (3) the more localised yellow form from the Ohrigstad district, with narrow, oblong, yellow racemes (the buds and flowers are all bright yellow). *Cryptopoda* means 'hidden foot', referring to the flower stalks which are hidden by the large leafy bracts. The Afrikaans common name is Geelaalwyn.

Eben van Wyk

Eben van Wyk

Eben van Wyk

Eben van Wyk

Eben van Wyk

Aloe gerstneri

Plants grow as single rounded rosettes. Stems are very short or absent. The leaves are long and relatively broad, up to 120 mm wide at the base, with both the upper and lower surfaces without thorns in mature plants (a few thorns may be present near the tip of the leaf on the lower side). Widely spaced sharp teeth are found along the leaf margins. These are characteristically pale brown and arise from a white base. Inflorescences are single in young plants but may have two or three branches in mature specimens. The racemes are up to 350 mm long and are not distinctly bicoloured, although the buds are usually a darker orange than the open flowers. Each raceme is long and narrow, and gradually tapers towards a narrow tip. The flowers are broad and tubular, about 30 mm long and bright orange.

Flowering occurs in February and March.

Aloe gerstneri differs from similar species such as *A. aculeata* and *A. reitzii* by the virtually spineless leaf surfaces of the mature plants and the pale brown marginal teeth which arise from a white base. It is also similar to *A. petricola* but the latter has longer, bicoloured racemes and it flowers in winter.

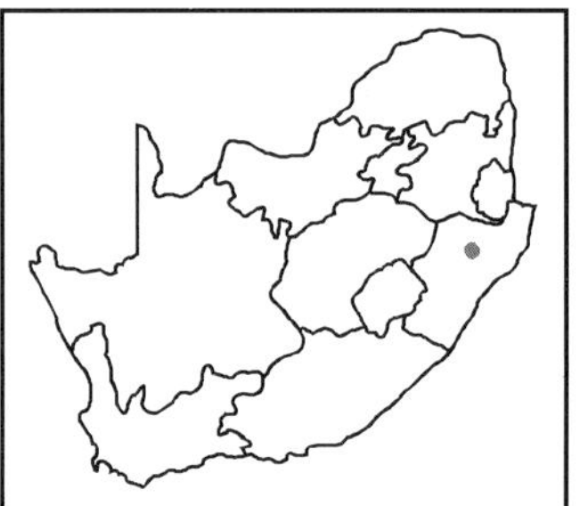

This species is found in a small area in northern KwaZulu-Natal, where it grows on rocky slopes and outcrops.

As a result of the localised distribution and decreasing numbers of plants, *A. gerstneri* is now considered to be critically rare.

Aloe gerstneri was named after the Rev. J. Gerstner who first discovered it in 1931. The only known common name is the Zulu word isiHlabane. The species is very attractive and grows well in cultivation.

Ben-Erik van Wyk

Ben-Erik van Wyk

NBI

Aloe glauca

Plants are stemless or may have short stems. The leaves are up to 400 mm long and do not have spots. They are distinctly blue-grey in colour, with faint longitudinal lines. The leaf margins have beautifully contrasting reddish-brown teeth. The lower leaf surfaces often have small, scattered spines towards the tips. The inflorescence is an unbranched raceme. Up to three cone-shaped racemes are borne one after the other from a single rosette. The peduncle is very robust and is covered by large leafy bracts. The buds on a developing raceme are similarly covered by large bracts and remain vertically disposed on the peduncle with only the open flowers becoming pendulous. Flower colour varies from pink to pale orange.

Flowering occurs from August to October.

Aloe glauca is readily distinguished by its fairly broad, blue leaves. The cone-shaped racemes with their robust peduncles also serve to characterise the species. Although *A. glauca* varies considerably across its distribution range, only one variety is recognised, namely *A. glauca* var. *muricata*. It differs from the typical variety in having greener and more spreading leaves. The lower leaf surfaces are more inclined to be tuberculate near the tip, and the marginal teeth are larger and more distinctly red in colour.

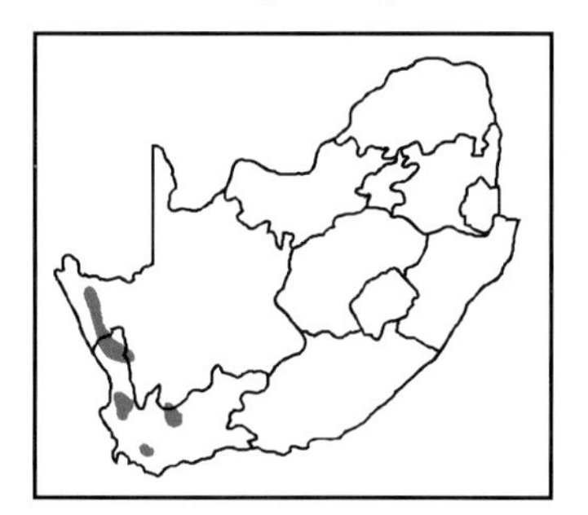

Aloe glauca occurs in the dry parts of the south-western Cape, where it prefers rocky hills and mountain slopes. The species is found from Swellendam in the south, northwards to Laingsburg in the Karoo and Steinkopf in Namaqualand.

The species is not threatened in its natural habitat.

Aloe glauca does not do well in cultivation in the summer rainfall region. The Afrikaans common name is Blouaalwyn, literally meaning 'blue aloe'. The Latin name *glauca* means 'grey', also referring to the greyish-blue leaves of this species.

Kobus Venter

Marike Bruwer

Pitta Joffe – NBI

Aloe globuligemma

Plants are more or less stemless or may form creeping stems of up to 1 m long. Numerous blue-green leaves with white edges are arranged in basal rosettes. The leaves are lance-shaped and are armed with numerous firm yet rather harmless white prickles. The leaves are erect but become recurved in the upper third. The inflorescence is branched and has horizontal branches. It carries upturned flowers not unlike the arrangement of the flowers of *A. marlothii*. The flower buds are red but turn a dull pinkish-white at maturity.

Aloe globuligemma flowers in July and August.

The distinctive flower shape makes this species easy to recognise, and it appears to have no close relatives. The leaves lack spots or lines and are uniformly blue-green. The small, club-shaped buds and flowers have a globular appearance due to the fact that they become much wider towards the mouth. (Differences between *A. globuligemma* and *A. chabaudii* are discussed under the latter species.)

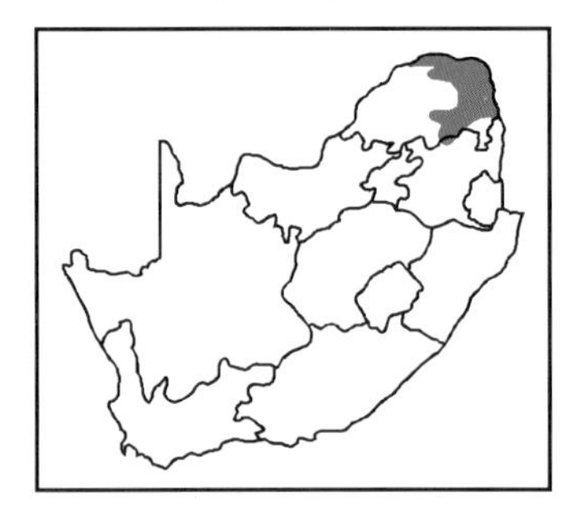

Aloe globuligemma occurs in warm, low-lying bushveld vegetation in the Northern Province, Mpumalanga and Zimbabwe.

Aloe globuligemma is not threatened anywhere in its distribution range.

The species makes an attractive addition to gardens in the summer rainfall area, especially because of its rather unusual bluish leaves with white margins. *A. globuligemma* will also produce basal suckers, although less proliferous than *A. chabaudii*. The descriptive Afrikaans common name Knoppiesaalwyn is derived from the distinctive shape of the flower buds. The Latin name is equally apt: *globuligemma* means 'globular bud'.

Eben van Wyk

Eben van Wyk

Eben van Wyk

Aloe haemanthifolia

This unusual and interesting species is stemless and forms dense clumps of up to twenty fan-like rosettes. The leaves are broad and strap-shaped, up to 200 mm long and 80 mm wide. They have bright green surfaces, pinkish margins and no spots. The leaf bases are sheathing and overlapping and the leaf clusters form characteristic fans, due to the opposite arrangement of the leaves. Flowers are borne in single racemes of up to 500 mm high. Each head-shaped raceme bears up to thirty tubular, slightly club-shaped, orange-red or scarlet flowers.

Flowering occurs mainly in October, but flowers may be found from September to December.

Aloe haemanthifolia is a very distinct species, not likely to be confused with any other. The fans of leaves resemble those of *A. plicatilis*, but the latter is a small tree with a definite trunk and branches. Another interesting difference is the leaf juice, which remains clear and watery in *A. haemanthifolia* but solidifies to a thick, pale, yellow waxy substance in *A. plicatilis*.

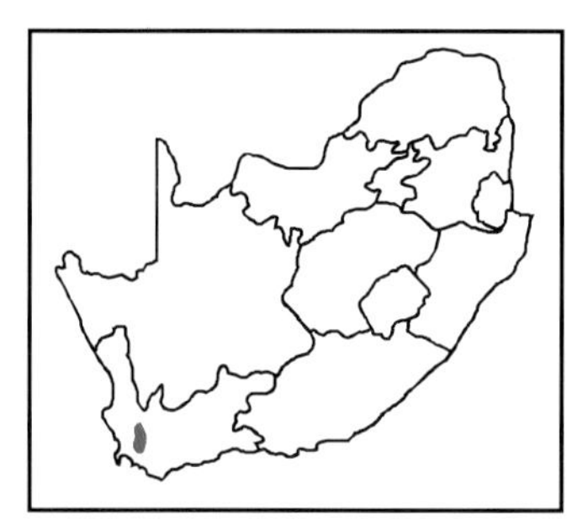

The species has a very limited distribution in the mountains of the south-western Cape, and is restricted to high mountain peaks, from Stellenbosch in the south to near Ceres in the north. The plants grow on steep rocky slopes (usually south-facing slopes) in areas of high winter rainfall.

Aloe haemanthifolia is a critically rare species but the inaccessible habitat in which it grows will hopefully contribute to its continued survival.

The scientific name refers to the superficial similarity of the leaves to those of *Haemanthus* (a well-known genus of bulbous plants). *Aloe haemanthifolia* is difficult to cultivate, even from seed, and unnecessary damage is done if plants are removed.

Kobus Venter

Kobus Venter

Edgar Swart

Aloe hereroensis

The plants usually grow as single rosettes but are sometimes divided into two or three heads. Stems are seemingly absent because they grow horizontally along the ground and are therefore not clearly visible. The greyish-green leaves are long and relatively narrow, with the tips curved inwards, giving the rosette a rounded appearance. The upper leaf surface has no spots but is marked with faint longitudinal lines; the lower surface may be sparsely to densely marked with H-shaped white spots, particularly in young plants. Along the leaf margin sharp, reddish-brown triangular teeth are present, but the upper and lower leaf surfaces have no spines or prickles. The inflorescence is repeatedly branched to form a wide cluster of ten to twenty or more racemes, all of which are more or less at the same level. The racemes are densely head-shaped, and are characteristically much wider than long and also uniform in colour throughout. The flowers are tubular and about 30 mm long. They are slightly inflated near the middle but become distinctly narrower towards the slightly upturned mouth. The flower colour is mostly orange-red, but various shades of red, orange and yellow are also quite common.

The flowering time is from June to September.

Aloe hereroensis may easily be distinguished from other stemless aloes by the spotted lower leaf surface (visible at least in young plants), the much-branched inflorescences and particularly by the short, broad racemes and distinctive shape of the flowers. It is superficially similar to the spotted aloes, but the upper leaf surface is without spots and the flowers do not have the characteristically rounded basal swelling.

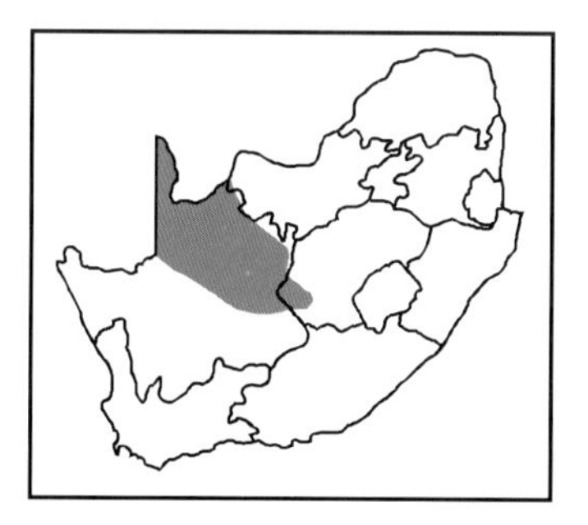

The species is widely distributed in the dry interior of South Africa (Northern Cape and Free State) and northwards to Namibia and Angola. It grows in flat sandy or stony places, or sometimes on rocky slopes.

Aloe hereroensis is not threatened.

This attractive aloe has no obvious relatives in South Africa. It may be grown in well-drained, alkaline soil. The common names are Sandaalwyn or Vlakte-aalwyn. The plant was originally found in northern Namibia, the home of the Herero, hence the scientific name.

Geoff Tribe

J.L.C.Marais

Pitta Joffe – NBI

Aloe krapohliana

This distinctive little aloe usually grows as a single rosette of up to 200 mm in diameter, but there is a form of the species with dense groups of up to fifteen rosettes. Stems are absent or inconspicuous except in very old specimens. The leaves are narrow and oblong, up to 200 mm long and up to 60 mm wide at the base. They are grey-green and without any spots or spines on the upper or lower surfaces, but the margins have minute white teeth. A characteristic feature is the peculiar greyish-brown transverse bands on the leaves, which give them a striped appearance. The inflorescences are simple or sometimes two-branched, and up to six of them may appear from each rosette. The dense, oblong racemes are remarkably large for such a small plant and are up to 150 mm long and 60 mm in diameter. The flowers are tubular, about 35 mm long and dull red with the tips greenish-yellow.

The flowering period is from June to August.

Aloe krapohliana is a unique species, not likely to be confused with any other. It may be identified by the small size of the rosettes, the remarkably large racemes (for such a small plant), and the banded leaves. The only other species which occasionally has banded leaves is the much taller *A. cryptopoda*.

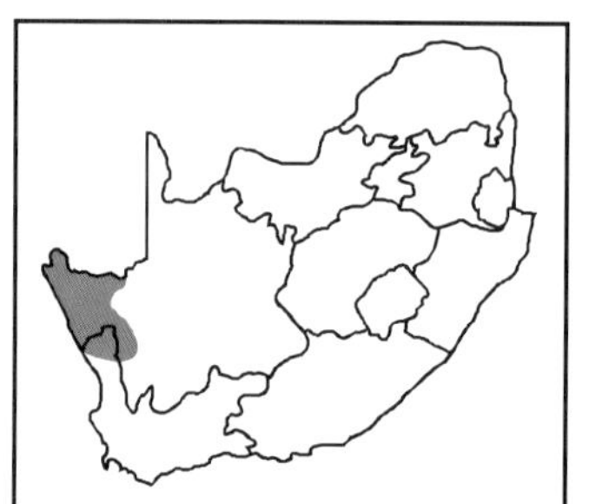

The species grows on sandy flats and rocky slopes in the extremely arid north-western corner of the Cape.

Aloe krapohliana is vulnerable as a result of collection, overgrazing and habitat destruction by mining activities.

This beautiful aloe is unfortunately difficult to grow and rarely survives for more than a few years outside its natural habitat. It was named after the first collector, H.C. Krapohl. A form of the species with small rosettes in clusters of up to fifteen heads has recently been described as *A. krapohliana* var. *dumoulinii*.

Graham Williamson

Forsyth van Nierop

Pitta Joffe – NBI

Aloe lutescens

Plants grow as small dense groups of rosettes arising from a short horizontal stem. The leaves are narrow and oblong, yellowish-green in colour, with the tips curved inwards. Small sharp teeth occur along the margin, but the upper and lower surfaces are without thorns or prickles. The inflorescences divide into three branches, each with a narrow, oblong raceme of about 400 mm long and up to 70 mm in diameter at the base. The racemes are distinctly bicoloured, with the buds dark red and the open flowers bright yellow.

The species flowers in June and July.

Aloe lutescens is very similar to *A. cryptopoda* but may be distinguished by the densely clustered growth habit, the yellowish-green leaves and the narrow, oblong inflorescences. When in flower, it may easily be confused with the bicoloured form of *A. cryptopoda*, previously known as *A. wickensii*. However, this form has shorter, broader, conical racemes and the flowers are slightly upturned at the mouth.

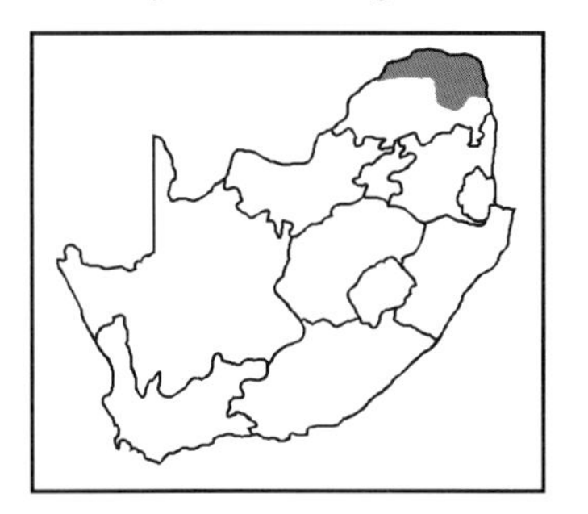

The species grows on dry rocky slopes and stony places in savannah vegetation and is often found in shade or partial shade. It is more or less restricted to the area between the Soutpansberg and the Limpopo River. Since *A. cryptopoda* has a more southern distribution in South Africa, natural populations of the two species are not likely to be confused.

Aloe lutescens is not threatened.

In the western extreme of the distribution area, plants are found which appear to be intermediate between *A. lutescens* and *A. cryptopoda*. The two species are undoubtedly very closely related, but they are generally easily distinguished. As a garden plant, *A. lutescens* grows well in frost-free areas but it is less showy than *A. cryptopoda*. The scientific name was well chosen: *lutescens* means 'becoming yellow', referring to the red buds which become yellow when the flowers open.

Eben van Wyk

Eben van Wyk

Eben van Wyk

Aloe melanacantha

Plants grow as single rosettes or more often in groups of up to ten or more dense, ball-shaped rosettes. Stems are short and inconspicuous, even in old specimens. The leaves are narrow and triangular, about 200 mm long and 40 mm wide at the base. They curve gracefully upwards and inwards, giving the rosettes a neat, rounded appearance. Unusual characteristics of the leaves also include the brownish-green colour, the firm texture and rough surface, and particularly the large black thorns along the margins and keel. Towards the base of the leaf the thorns may be relatively short, with a whitish colour, but those on the upper parts are up to 10 mm long and distinctly black in colour. The inflorescences are usually simple, 1 m high, with a single oblong rosette of about 200 mm by 80 mm. The flowers are tubular and bright red, but turn yellow after opening.

The species flowers in May and June.

This very distinctive aloe is easily identified by the rounded, ball-shaped rosettes, the firm brownish-green leaves and particularly by the long black thorns on the leaf margins and keel.

The plants grow in sandy or rocky areas and prefer arid conditions. The distribution extends from Nieuwoudtville and Bitterfontein northwards to southern Namibia.

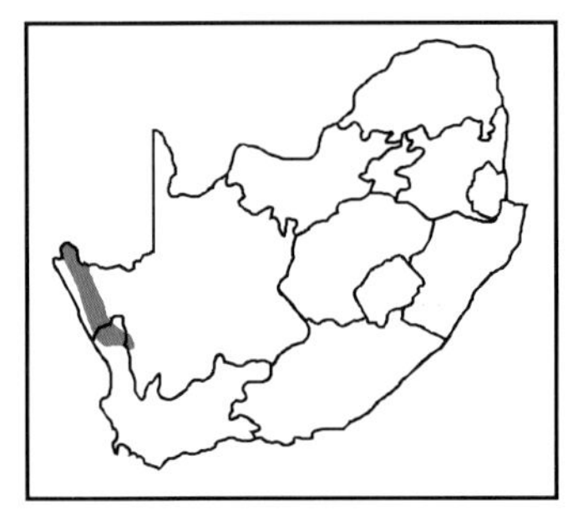

Aloe melanacantha is not threatened.

The scientific name *melanacantha* ('black thorns') was well chosen because it highlights the most conspicuous feature of this interesting aloe. Common names include Kleinbergaalwyn and Goree.

Collecting this aloe is wasteful and causes unnecessary damage, because the plants do not thrive out of their natural habitat.

Kobus Venter

Ben-Erik van Wyk

Pitta Joffe – NBI

Aloe peglerae

Plants grow as solitary rosettes or rarely form small groups. Stems are very short or absent. The leaves curve inwards, resulting in a neat, rounded head. Each leaf is narrow and triangular, greyish-green or reddish-green (during the dry months), about 250 mm long and up to 80 mm wide at the base. In addition to the 5 mm long, sharp, brown teeth along the margins, there are also two short rows of spines along the middle of both the upper and lower surfaces. The remarkably short inflorescence is usually simple and unbranched, with a robust, bicoloured raceme. The flowers are broad and tubular, and up to 30 mm long. Each flower is dull red in the bud stage but becomes pale greenish-yellow when it opens, with conspicuous dark purple stamens protruding for some 25 mm from the mouth of the tube.

Aloe peglerae flowers in July and August.

The single, dense racemes, which are borne directly above the rosettes, are exceptionally large for such a small plant. This unique appearance makes it unlikely that *A. peglerae* will be confused with any other species. When not in flower, it is somewhat similar to *A. aculeata* but the latter has numerous tuberculate thorns spread over the entire lower (outer) surface of the leaf.

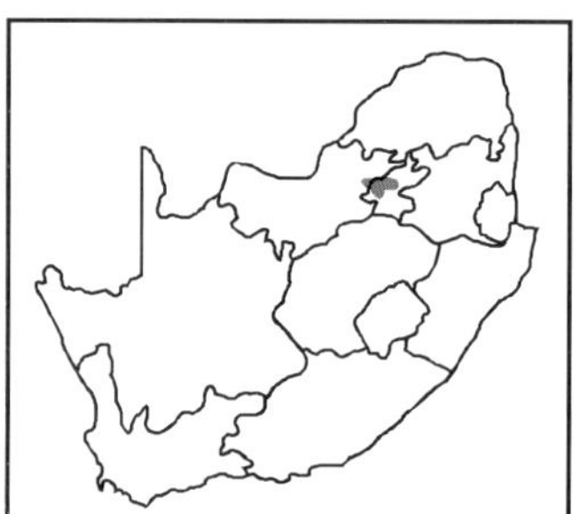

Aloe peglerae is found mainly on the Magaliesberg and Witwatersberg, from Pretoria westwards to Rustenburg and southwards to Krugersdorp. It grows in rocky areas, usually on northern slopes.

As a result of the localised distribution near densely populated areas, the species has become critically rare.

This striking aloe with its peculiar stunted appearance was named after a well-known early plant collector, Alice Pegler. The common names Red-hot poker and Bergaalwyn have been recorded. It is relatively easy to cultivate if given a sheltered position and well-drained soil.

Forsyth van Nierop

Ben-Erik van Wyk

Aloe petricola

Plants form single, densely leaved rosettes. Stems are absent or very short and inconspicuous. The greyish-green leaves are long, relatively broad at the base and narrow at the tip, with the upper parts distinctly curved inwards to give the rosette a rounded appearance. A few scattered thorns are sometimes present on the upper and particularly the lower surfaces. Sharp brown triangular teeth of about 5 mm long occur along the leaf margins. The inflorescences are single in young plants or divided into three or four (occasionally up to six) branches in old plants. The racemes are long, narrow, very densely flowered and usually distinctly bicoloured. The flowers are tubular, somewhat wider in the middle and up to 30 mm long. In the common bicoloured forms, the buds are red and the open flowers greenish-white; in some plants the buds may be orange and the open flowers yellow.

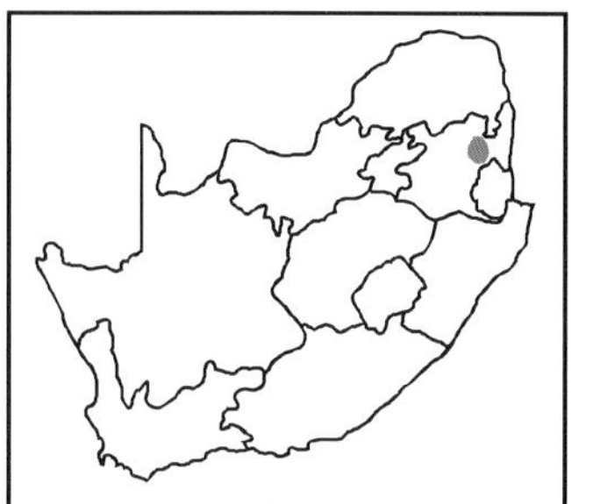

The flowering time is from July to August.

Aloe petricola is similar to *A. aculeata* but it lacks the abundant tuberculate spines on the leaf surfaces, which are so typical of the latter. When in flower, the species can easily be identified by the densely flowered, distinctly bicoloured racemes.

The species has a restricted distribution around Nelspruit in Mpumalanga, where it is fairly common on sandstone slopes and granite outcrops. It usually grows in large groups and is often found in very shallow soil.

Aloe petricola is not threatened.

The Latin name of the species was well chosen: *petricola* means 'inhabitant of rocky places'. This aloe grows well in cultivation and is exceptionally beautiful when in full flower.

Ben-Erik van Wyk

Ben-Erik van Wyk

Trevor Coleman

Aloe polyphylla

This high-altitude species from Lesotho is one of the most spectacular plants of Southern Africa. The plant usually forms a single rounded head but occasionally divides into two or three rosettes. The most striking feature is the distinctive clockwise or anticlockwise spiral arrangement of the leaves, which gives the plant a peculiar symmetrical appearance. The leaves are broad, up to 300 mm by 100 mm, with a grey-green colour. Other distinctive characteristics are the raised edge or keel along the lower leaf surface of young plants, which is invariably off-centre, that is not exactly in the middle of the leaf; and also the leaf tip, which becomes dry and turns purplish-black. The inflorescence is up to 600 mm high and is branched low down on the plant into three to eight racemes. Each raceme is densely flowered, with tubular, pale red to salmon-coloured flowers of up to 55 mm long.

Flowering occurs mainly in September and October.

The spiral aloe is a truly unique species and the spiral arrangement of the leaves makes it easy to identify.

Aloe polyphylla is restricted to steep basaltic mountain slopes in Lesotho, at altitudes above 2000 m. Moisture from clouds and mist adds substantially to the annual rainfall, which can be over 1000 mm. The plants are often under snow in winter.

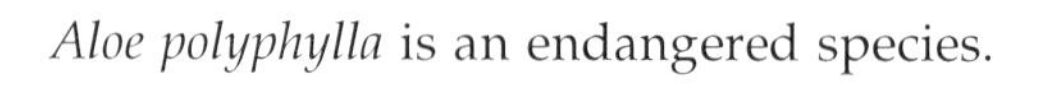

Aloe polyphylla is an endangered species.

Despite being protected by law, numbers have declined rapidly in recent years as a result of unscrupulous collectors who sell the plants to local people and visitors. This is done in spite of the fact that there is virtually no chance of survival outside the natural habitat. Soil and moisture requirements are simply too specialised for the plants to last for more than a few years in cultivation. The protected status of the plant unfortunately only seems to make it more desirable. Dam building activities do not appear to affect the plants directly, but many plants are now removed from remote areas as a result of the increased accessibility. *Polyphylla* literally means 'many leaves'. The Afrikaans common name is Kroonaalwyn.

Duncan Butchart

Bill van der Riet

Godfrey Symons

Aloe pratensis

Plants usually branch at the base to form small groups of up to six, with each head about 200 mm in diameter. Solitary plants are rare. The leaves are neatly arranged in regular, densely leaved rosettes. Each leaf is relatively broad, up to 150 mm by 50 mm, grey in colour and with more or less distinct longitudinal lines. The upper surface is usually without tubercles or spines, but the lower surface (particularly the midrib and keel) often has several brown spines arising from a white tuberculate base. Along the margins of the leaf there are sharp triangular spines of up to 5 mm long. These are reddish-brown in colour but arise from a white base. Up to four inflorescences may arise from the rosette. Each one is unbranched and about 600 mm high, with large, striped, papery bracts along the entire length of the peduncle. The raceme is about 200 mm long and 100 mm wide, with the buds densely crowded and covered by the bracts. The flowers are pinkish-red and up to 40 mm long. The petals are free to the base but form a cylindrical tube which is only very slightly widened in the middle. The stamens and style protrude only slightly from the mouth of the tube. Mature fruits and flowers are found together on the same raceme.

The flowering time varies considerably, from June to October.

Aloe pratensis can easily be distinguished by the bracts on the peduncles and especially by the large white tubercles at the base of the spines and marginal teeth.

The distribution area stretches from sea level at Grahamstown in the south to the high altitude of Cathedral Peak and Champagne Castle on the eastern side of the Drakensberg. Extreme values for the annual rainfall are 500 mm in the Eastern Cape Province, and over 1 000 mm in highlying areas of KwaZulu-Natal.

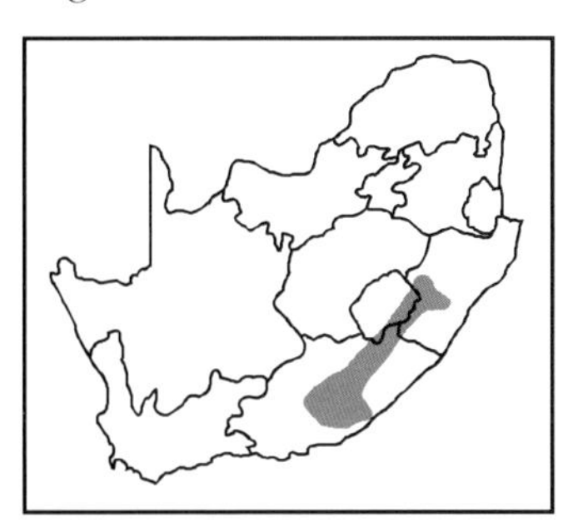

The conservation status of this species is insufficiently known. Although fairly common at some localities, it has come under increasing pressure due to agricultural development and collection.

The name *pratensis* ('growing in a meadow') is not very appropriate, as the species is invariably found in rocky places, wedged in amongst rocks. *Aloe pratensis* does not seem to thrive in cultivation and is rarely grown in gardens.

Cameron McMaster

Aloe reitzii

The plants usually grow as single rosettes but may rarely divide into two or three heads. Stems are absent or very short and inconspicuous. The leaves are long and relatively broad, up to 120 mm wide at the base. The upper surfaces of the leaves are without thorns in mature plants, but have a short line of brown thorns along the lower side near the tip of the leaf. Sharp reddish-brown teeth of about 3 mm long are present along the leaf margins. Inflorescences have a single raceme in young plants but there are between two and six branches in mature specimens, each with a bright red raceme of about 400 mm long. The flowers are dark red but turn yellowish with age. They are up to 50 mm long, narrow, tubular and distinctly curved.

The species usually flowers in February and March, but a distinct winter-flowering form is also known (see below).

Aloe reitzii may be distinguished from similar species (*A. aculeata*, *A. gerstneri* and *A. petricola*) by the relatively long and tubular flowers that are distinctly curved. The flowers point downwards and are pressed against the stalk, with the outer (exposed) part of the tube being dark red and the inner (lower) part of the tube lemon yellow. As a result, the individual flowers are characteristically bicoloured in this species. The racemes, however, are not markedly bicoloured as in *A. petricola*.

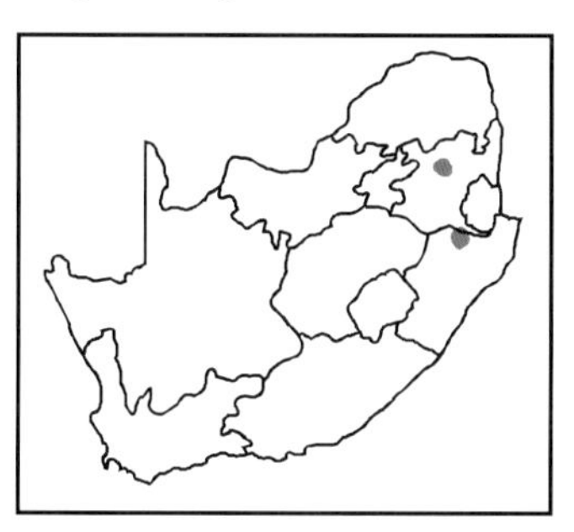

The known distribution of the species is limited to small areas in the Belfast district in Mpumalanga and in northern KwaZulu-Natal. It grows on rocky slopes in grassland.

The conservation status of *Aloe reitzii* var. *reitzii* is considered to be indeterminate, but *A. reitzii* var. *vernalis* (see below), has become critically rare as a result of the very localised distribution.

This distinctive and spectacular species grows well in cultivation and deserves to be propagated and planted more often. It was named after F.W. Reitz, of the well-known South African Reitz family. The plants usually flower in February and March, but a winter-flowering form has been described as a distinct variety, namely *A. reitzii* var. *vernalis*.

Gawie Dednam

Gawie Dednam

Gawie Dednam

Aloe reynoldsii

Plants are stemless or have short stems covered by the remains of dead leaves. Multiheaded specimens are often encountered in habitat. The leaves are an attractive pale bluish-green to yellow colour, contrasting strikingly with the minutely toothed pinkish-red leaf borders. Both surfaces have numerous H-shaped whitish spots that coalesce to form longitudinal lines. The branched inflorescence consists of numerous racemes. The flowers are yellow and have slight basal swellings.

Aloe reynoldsii flowers in September and October.

The species can be easily distinguished from the closely related *A. striata* by its more or less wavy leaf margins and small teeth along the leaf margin. It characteristically has sparse racemes with yellow flowers. It is in most respects a more dainty plant than *A. striata. Aloe reynoldsii* is essentially a cliff dweller and is more likely to form large clumps by division.

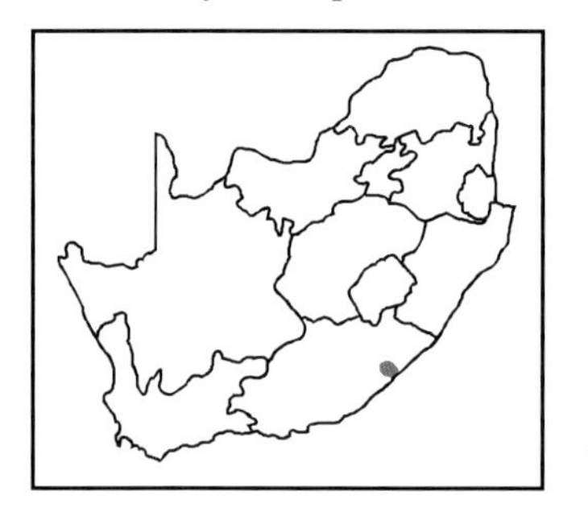

The species occurs on cliffs in a humid eastern part of the Eastern Cape Province. It is restricted to an area around the mouth of the Bashee River.

Aloe reynoldsii is included in the vulnerable conservation category.

The scientific name honours the well-known aloe expert, Gilbert W. Reynolds. The species is attractive in cultivation and will tolerate widely differing watering schedules, and even some light frost.

Ben-Erik van Wyk

Ben-Erik van Wyk

Ben-Erik van Wyk

Aloe striata

Plants are usually stemless, but very old specimens may form a short, creeping stem covered with persistent dried leaves. The light blue-green leaves are flat and broad and distinctly boat-shaped. The margins are spineless and have soft, reddish edges. Up to three branched inflorescences are formed. The shape of the racemes varies from head-shaped to cone-shaped. The flowers have slight basal swellings, to some extent resembling the flowers of the spotted aloes. Flower colour varies from pinkish-red to bright orange, and a yellow form has been recorded from the Eastern Cape Province.

The typical *A. striata* flowers from July to October. The two other subspecies (see below) have different flowering times: subsp. *karasbergensis* flowers from January to March and subsp. *komaggasensis* in December and January.

Aloe striata subsp. *striata* is very distinctive by virtue of its smooth-edged, pale greyish-green leaves. The leaves of subsp. *karasbergensis* are a brownish colour with distinct longitudinal lines and whitish margins, while subsp. *komaggasensis* has white-edged leaves with only faint lines. These three entities, which are nowadays included in a wide concept of *A. striata*, are very distinctive and cannot be confused with any other aloe species.

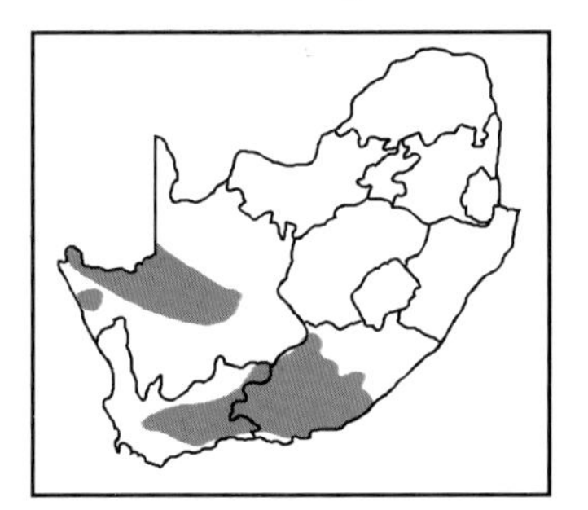

Aloe striata subsp. *striata* is widely distributed in the dry parts of the Western and Eastern Cape Provinces. Subsp. *karasbergensis* occurs in the Northern Cape Province, from Prieska to northern Namaqualand, and in southern Namibia. Subsp. *komagga-sensis* has a more restricted distribution in the mountains around Komaggas in Namaqualand.

Of the three subspecies of *Aloe striata*, only subsp. *komaggasensis* is regarded as threatened. It is treated as rare due to its localised geographical distribution range, overgrazing and injudicious collection. *Aloe striata* subsp. *karasbergensis* is currently listed as not threatened.

Aloe striata is widely cultivated as a garden ornamental and grows easily under a variety of environmental conditions. Even subsp. *karasbergensis*, which grows naturally in one of the most arid regions of Southern Africa, will tolerate a fairly high rainfall in cultivation. Subsp. *komaggasensis* has not yet been widely introduced into cultivation. *Striata* means 'marked with lines'. The common names of *A. striata* subsp. *striata* are Blouaalwyn, Gladdeblaaraalwyn, Makaalwyn, Streepaalwyn and Vaalblaaraalwyn; and Coral aloe.

Geoff Tribe

Ben-Erik van Wyk

Ben-Erik van Wyk

Aloe suprafoliata

Plants are solitary and stemless. The rosettes of mature plants have an indistinct spiral twist, not unlike the rosettes of *A. pluridens*. In young plants the leaves are two-ranked and only take on the spiral arrangement after a few years. The leaves are bluish-green and have sharp marginal teeth. The leaf tips are often tinted red. Up to three unbranched racemes are produced simultaneously. Flowers are pinkish-red, cylindrical and pencil-shaped.

Aloe suprafoliata flowers from May to July.

When not in flower, the species may be confused with *A. pretoriensis*. However, the leaves of *A. pretoriensis* are more inclined to die back from the tips, and their terminal portions are usually bright red. When in flower, *A. suprafoliata* can easily be distinguished by its much shorter, unbranched inflorescences, its narrow racemes which have a silvery sheen, and the large, rounded flower bracts.

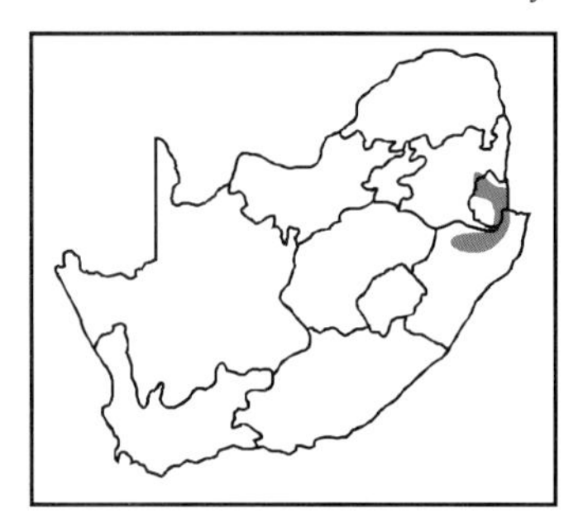

The species grows on rocky slopes in Mpumalanga, northern KwaZulu-Natal and Swaziland. In its habitat *A. suprafoliata* is often subjected to low temperatures and mist.

The species is not threatened.

The leaves of young plants of *A. suprafoliata* are distinctly arranged in two flat or recurved rows, which have given rise to the very descriptive Afrikaans common name Boekaalwyn. The Latin name *suprafoliata* also refers to the leaves of young plants that are seemingly situated on top of each other in two rows, resembling the pages of an open book. The species is fairly easy to cultivate, and with its blue-green leaves and long red flowers makes an attractive addition to any garden.

GW Reynolds – NBI

Frits van Oudtshoorn

Frits van Oudtshoorn

Pitta Joffe – NBI

Aloe thorncroftii

Plants never form suckers and always occur as solitary specimens. Stems are usually absent or very short. The broad, pointed leaves are up to 400 mm long, with the terminal part dried and shrivelled. Leaf colour is somewhat variable, ranging from dark reddish-green to bluish or greyish-green, with the dried portion distinctly red-tinted. The leaf margins have sharp, brownish, triangular teeth. The simple inflorescences are sparsely flowered and up to 500 mm high. It is not uncommon to find plants producing two racemes simultaneously from a single rosette. The flowers are an attractive pale or dark red and up to 60 mm long.

Aloe thorncroftii usually flowers in September.

The species is most closely related to *A. suprafoliata*, from which it differs in having longer flowers, more sparsely flowered racemes and broader leaves.

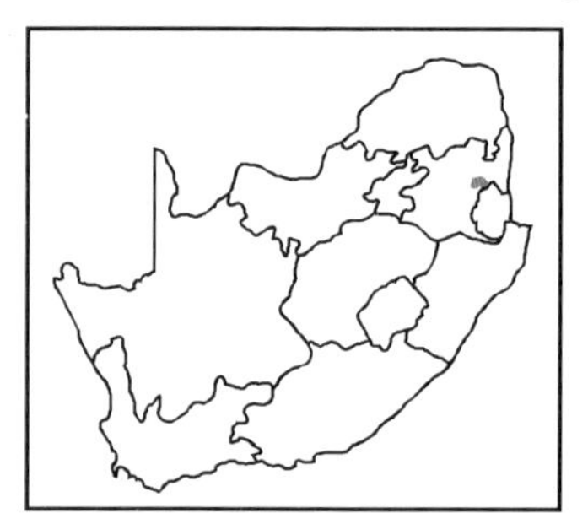

Aloe thorncroftii is a high-altitude species, restricted to a small area in the mountains around Barberton in Mpumalanga. Plants often grow almost horizontally from steep krantzes and crevices, with the inflorescences curving gracefully upwards.

Due to its small world population *Aloe thorncroftii* is regarded as rare.

It is more difficult to cultivate than the closely related *A. suprafoliata* and is not often encountered in collections. The species was first collected by George Thorncroft, after whom it was named.

GW Reynolds – NBI

Ben-Erik van Wyk

Ernst van Jaarsveld – NBI

SPECKLED ALOES

Group 7

All the species included here are closely related, so that the speckled aloes are a natural group. The growth form varies greatly and the plants may be single-stemmed or multistemmed. Species of the group are distinguished by the presence of **at least a short stem** and by **numerous small spots on the upper and lower leaf surfaces**. Small spots may be found on several *Aloe* species, and it is therefore necessary to consider other characteristics as well. The flowers are tubular and not basally inflated as in the spotted aloes, and the leaves are comparatively longer and narrower than those of the spotted aloes. Another feature of the speckled aloes is their leaf colour, which often turns distinctly reddish-brown, even when the plants are comparatively well watered. The species all have a **western distribution** in South Africa (southern to western parts of the Cape), so that the origin of a plant is a particularly useful clue in deciding whether it belongs here or not.

SPECIES

A. framesii
A. gariepensis
A. khamiesensis
A. microstigma
A. pictifolia

Ben-Erik van Wyk

Aloe microstigma

Aloe framesii

Plants form dense groups of up to twenty rosettes. The branched stems are horizontal and inconspicuous. The leaves are long and relatively narrow, about 300 mm long and 70 mm wide at the base. Numerous small white spots are usually present on both surfaces. There are no thorns or prickles except for reddish-brown triangular teeth along the margins. The inflorescences are single or mostly two to three-branched and up to 800 mm high, with oblong, pointed racemes. The flowers are tubular and usually orange-red with the tips greenish-yellow.

Flowering occurs in June and July.

Aloe framesii is very closely related to *A. khamiesensis* and differs mainly in the densely crowded groups of rosettes, the absence of erect stems and the smaller number of racemes in each inflorescence (the inflorescence usually has up to three branches in *A. framesii* but more than four branches in *A. khamiesensis*). It is a coastal species, while *A. khamiesensis* occurs in the mountains.

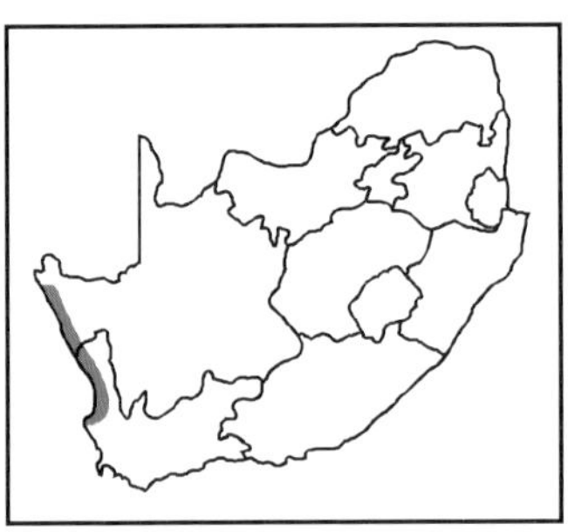

The natural distribution area is along the west coast of the Cape, from Saldanha northwards to near Port Nolloth. The plants are usually found in flat, sandy places.

Aloe framesii is not threatened.

This attractive aloe was named after Percy Ross Frames, who first collected the plant in Namaqualand, just north of Port Nolloth. Plants do not grow well in gardens and should therefore not be removed from their natural habitat.

GW Reynolds – NBI

Kobus Venter

Pitta Joffe – NBI

Aloe gariepensis

Plants are usually solitary. They vary in size, from small and stemless to larger, up to 1 m high, with a short erect stem. The leaves have a characteristic striped appearance due to numerous longitudinal lines. The leaves are copiously spotted on both surfaces in young plants, and some spots usually remain on the upper surfaces when the plant matures. Small, sharp, triangular teeth occur along the brown horn-like leaf margins, but there are no thorns on the leaf surfaces. The inflorescence is simple (unbranched) and up to 1,2 m high, with a single, narrow, oblong raceme. The racemes are yellow to greenish-yellow, or sometimes the buds are red and the open flowers yellow, giving a bicoloured effect. The flowers are relatively short, tubular and slightly wider towards the mouth. The young buds are more or less hidden by long bracts.

The flowering time varies from early July to September.

This aloe can easily be recognised by the long narrow racemes and the long flower bracts. The leaves are yellowish-green and often turn pink or bright reddish-brown during dry periods. The lines on the leaves are also useful to distinguish the species from its close relatives. (Differences between *A. gariepensis* and the highly localised *A. chlorantha* are given under the latter.)

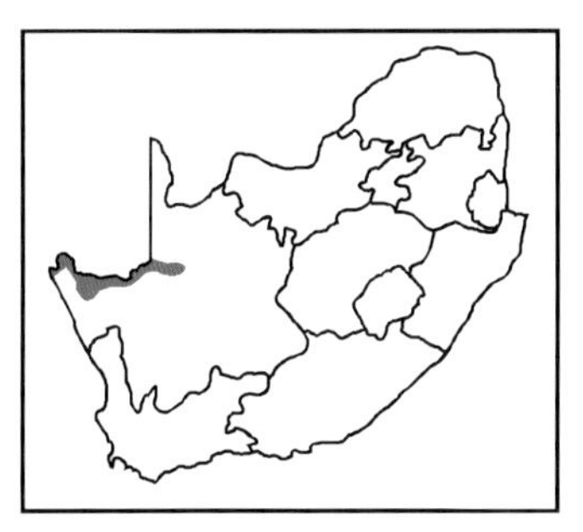

In South Africa, *A. gariepensis* is restricted to the northern boundary of the Northern Cape Province, from Keimoes westwards towards northern Namaqualand. In this arid environment it grows on rocky slopes or in rock crevices.

The species still occurs in fairly large numbers and is not threatened.

Aloe gariepensis is appropriately named, because the natural distribution closely follows the lower reaches of the Gariep River (nowadays better known as the Orange River). The species hardly ever thrives in cultivation unless specialised care can be given.

Forsyth van Nierop

Graham Williamson

Ben-Erik van Wyk

Ben-Erik van Wyk

Pitta Joffe – NBI

Aloe khamiesensis

Plants are erect and up to 3 m tall, usually single-stemmed with only one rosette, but sometimes the stem is branched into two. The leaves are long and relatively narrow, about 400 mm long and 80 mm wide at the base. The lower parts of the leaves curve upwards and the tips often curve outwards, giving a graceful appearance to the plants. Small white spots are usually present on the upper surfaces and particularly the lower surfaces. Reddish-brown triangular teeth are present along the margins. The inflorescences are repeatedly branched to form between four and eight racemes. The broadly triangular racemes are about 300 mm long, with tubular, orange-red flowers of which the tips are greenish-yellow.

The flowering time is in June and July.

In general appearance, the leaves and flowers of this species are identical to those of the closely related *A. framesii*. It differs, however, in the single erect stems and more numerous racemes (up to eight) in each inflorescence. Furthermore, *A. khamiesensis* has a more inland distribution at higher altitudes, while *A. framesii* occurs at low altitudes along the coast.

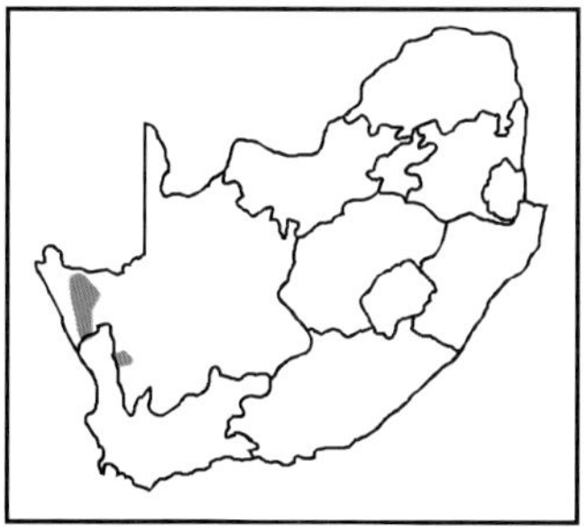

The species has a narrow distribution in the Northern Cape Province and is found only in the mountainous part of Namaqualand and at isolated localities in the Calvinia district.

The conservation status of *A. khamiesensis* is considered to be insufficiently known. At some localities, the numbers of plants have declined markedly as a result of unscrupulous collection.

The scientific name refers to the Khamiesberg and Khamieskroon, where the plants were originally collected. Common names that have been recorded include Tweederly, Aloeboom and Wilde-aalwyn. This beautiful aloe unfortunately does not grow well in gardens and should best be enjoyed in its natural habitat. The national tree number is 29.3.

Kobus Venter

Colla Swart

Colla Swart

Aloe microstigma

Plants usually occur as single rosettes but sometimes form small groups. The stems are short, but may be up to 500 mm long in old plants. The leaves are long and relatively narrow, about 300 mm long and 60 mm wide at the base. They are usually reddish-green and give a distinct reddish appearance to the plants, particularly in the dry season. Numerous white spots are usually present on the upper and lower sides of the leaves. Sharp, reddish-brown triangular teeth are present along the margins, but there are no spines or prickles on the leaf surface. The inflorescences are up to 1 m high and are always simple, with a single, narrow, oblong raceme. Two or three of them are usually produced simultaneously. The racemes are mostly bicoloured, with dull red buds turning yellow when the flowers open. In some places, however, the buds and open flowers may be uniformly yellow or red. The flowers are tubular and slightly swollen in the middle.

The flowering period is from May to July, with a peak period in June.

Aloe microstigma may be distinguished from *A. khamiesensis* and *A. framesii* by its smaller size, the longer, thinner racemes and the more densely spotted leaves. It is particularly similar to *A. gariepensis*, but the leaves are mostly copiously spotted in mature specimens and the flower bracts are much shorter, not hiding the buds. (Differences between *A. microstigma* and the closely related *A. pictifolia* are given under the latter.)

This species is widely distributed in the dry interior of the Western and Eastern Cape Provinces, and often occurs in vast numbers as one of the dominant plants in the landscape. It is found in a variety of habitats, in flat open areas, steep rocky slopes or amongst bushes.

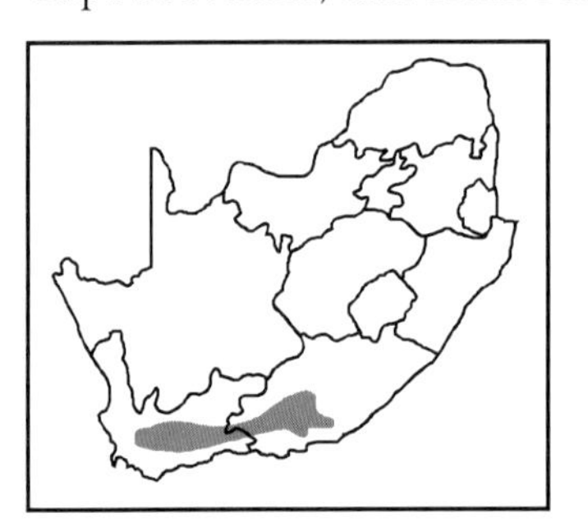

The species is exceptionally common and is not threatened.

The scientific name accurately describes the distinguishing feature of the plant (*microstigma* means 'very small spot'). It is surprising that there are no common names, given the abundance and wide distribution of this species. Plants grow well in cultivation, provided they are given a warm, sheltered place with well-drained soil.

Craig Hilton-Taylor

Geoff Tribe

Piet Vorster

Pitta Joffe – NBI

Aloe pictifolia

Plants usually form small groups of about three to seven rosettes from a short creeping or hanging stem. The reddish-green or pinkish leaves are long and very narrow, up to 150 mm long and no more than 25 mm wide at the base. Both leaf surfaces are copiously spotted with small, round, white spots and there are small reddish-brown teeth along the margins. The inflorescence is simple (i.e. unbranched) and up to 350 mm high, with small, narrow cone-shaped racemes. The tubular flowers are relatively small, up to 16 mm long and dull red in colour, with the mouth turning yellowish after opening.

The species flowers from July to September.

This close relative of *A. microstigma* can easily be distinguished by the much smaller and more clustered growth habit, the much narrower, ribbon-shaped leaves and the much smaller racemes and flowers.

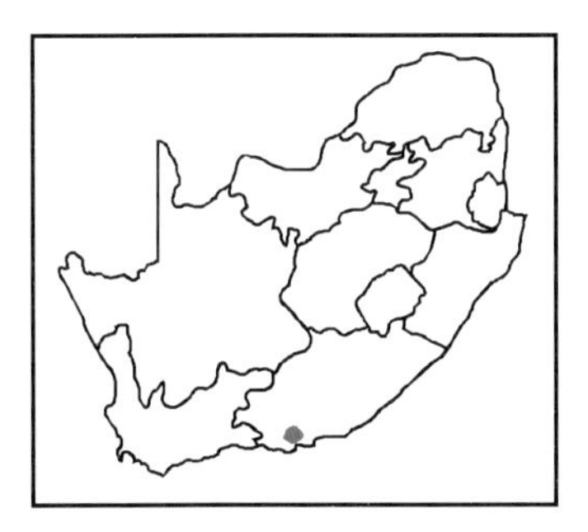

Aloe pictifolia has a restricted distribution and is known only from a small area north of Humansdorp in the Eastern Cape Province. It grows on steep rocky cliffs.

The species is not threatened, but the distribution is highly localised and care should be taken to protect the plant in its natural habitat.

This charming little aloe is relatively easy to cultivate and makes an attractive pot plant. The scientific name refers to the small spots on the leaves (*pictifolia* means 'painted leaves').

Gerhard Marx

Frits van Oudtshoorn

Nico Laubscher

Spotted aloes

Group 8

This is one of the most difficult groups in the genus *Aloe* because many of the species are not readily distinguished from one another. Although some of them, such as *A. grandidentata* with its clavate or club-shaped flowers, are quite distinct, most are exasperatingly difficult to identify, especially in the vegetative phase. The group is characterised mainly by the **copiously spotted leaves** which are borne in small to medium-sized rosettes. The **flowers** of most spotted aloes are bulbous, that is **conspicuously inflated at the base**, enlarging towards the throat. All the species are stemless or rarely have short stems. With the exception of *Aloe suffulta*, this is a natural assemblage of related species, often referred to as the maculate aloes. In general, the speckled aloes (Group 10), which are here justifiably recognised as a distinct entity, have at least a short stem, leaves with smaller spots and flowers that are not basally inflated.

SPECIES

A. affinis
A. branddraaiensis
A. burgersfortensis
A. dewetii
A. dyeri
A. fosteri
A. grandidentata
A. greatheadii
A. greenii
A. immaculata
A. lettyae
A. longibracteata
A. maculata
A. monotropa
A. mudenensis
A. parvibracteata
A. petrophila
A. prinslooi
A. pruinosa
A. simii
A. suffulta
A. swynnertonii
A. umfoloziensis
A. verdoorniae
A. vogtsii
A. zebrina

Pitta Joffe – NBI

Aloe greatheadii

Aloe affinis

Plants are usually solitary. The rosettes of the more or less unspotted, yellowish-green leaves are fairly large for a spotted aloe. The fleshy leaves are fairly broad at the base and tend to die back from the tip, giving them a distinctly triangular look. The leaves have dark brownish longitudinal lines on the upper surface. The leaf margins have horny edges and reddish teeth. Up to three inflorescences, each with between three and ten side-branches, are produced simultaneously or consecutively. These are fairly robust and specimens with inflorescences of up to 1 m high are common. The racemes are comparatively densely flowered and rounded or cone-shaped. The flowers are uniformly coloured but could be any shade of red, from pink to coral-red.

The species flowers in June and July.

The robust rosettes of the more or less unspotted, highly succulent leaves and the robust inflorescences carrying uniformly reddish flowers are distinguishing characteristics for *A. affinis*. The rosettes are rather flat and open, and the leaves take on a characteristic yellowish-green colour. Furthermore, the leaf edges have distinct brownish edges.

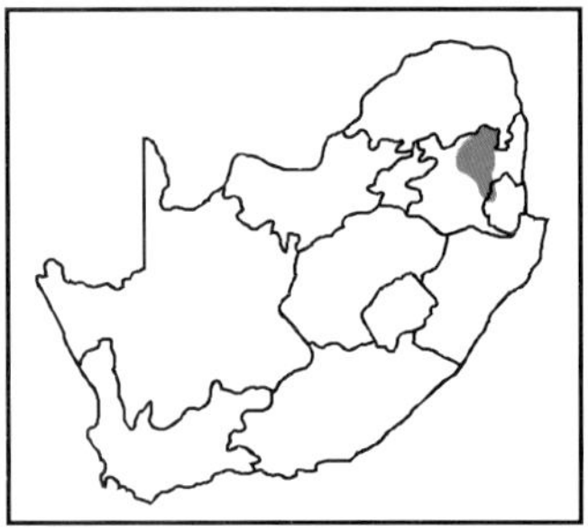

The species occurs in Mpumalanga, on rocky slopes in grassland and in the shade of bushes. It occurs from the Swaziland border northwards to Marieps Mountain. The eastern limit is Graskop and Pilgrim's Rest, and Lydenburg appears to be the furthest westerly distribution of the species. It is particularly common in Schoemans-kloof near Machadodorp.

The species is not threatened in its natural habitat, but large parts of its distribution range are nowadays planted with pines and eucalypts for commercial forestry purposes.

Aloe affinis is a rather attractive spotted aloe and does well in cultivation. However, its leaves are prone to black, rusty spots. *Affinis* means 'neighbouring' or 'allied to', referring to the close relationship between this species and other spotted aloes.

Pitta Joffe – NBI

Ben-Erik van Wyk

Pitta Joffe – NBI

Aloe branddraaiensis

Plants are stemless and form small clumps from basal suckers. On the upper leaf surfaces the species has more or less distinct stripes formed by the white spots which are distributed in longitudinal lines. In full sun the leaves take on a beautiful reddish hue, but lose much of this colouring in shady positions. The tips of the leaves tend to die back and are distinctly twisted. The leaf margins are armed with sharp brown teeth. Up to three much-branched inflorescences are produced simultaneously. Each of these can produce more than fifty head-shaped, loosely packed racemes. The flowers are coral-red.

The species flowers from June to July.

The head-shaped racemes and many side-branches produced by the inflorescences make this species fairly distinctive. The dark brown leaves with their whitish lines help to distinguish *A. branddraaiensis* from other spotted aloes.

In contrast to most spotted aloes, *A. branddraaiensis* has a restricted distribution in Mpumalanga where it occurs in the vicinity of Ohrigstad and northwards to the Olifants River. It grows on rocky hills and in the shade of bushes.

The species is not threatened.

Aloe branddraaiensis grows easily in frost-free areas. The scientific name is derived from Branddraai near Ohrigstad, the locality where it was first collected.

Pitta Joffe – NBI

Ben-Erik van Wyk

Pitta Joffe – NBI

Aloe burgersfortensis

Plants are usually solitary, but some specimens will occasionally form two or three heads from basal suckers. Leaf colour is very variable and ranges from dull brownish-green to bright green. The leaves vary from densely to weakly spotted. The inflorescence, once or twice branched, is sparsely flowered with flowers not unlike those of *A. zebrina*. Flower colour is most commonly a dull pinkish-red.

Flowering is from June to July.

This is yet another species of spotted aloe that hardly has any distinguishing characteristics, apart from its geographical distribution range and its immense variability! Unlike *A. zebrina*, the species flowers in winter.

The species grows in sandy soil in the open or in the shade of trees. It is a typical component of bushveld vegetation around Burgersfort and Steelpoort, and also occurs further to the south near Barberton.

Aloe burgersfortensis is not threatened.

There is little of interest to add to the description given above. Like most spotted aloes, *A. burgersfortensis* is immensely variable and a critical study of the group might well prove it to be the same as one of the other widespread species. *Aloe burgersfortensis* is easy to cultivate in the summer rainfall regions of South Africa. The species is named after the town of Burgersfort, where it was first collected.

Eben van Wyk

Eben van Wyk

Gideon Smith

Aloe dewetii

Plants are very large, stemless and do not form suckers. Individual rosettes can be more than 1 m in diameter. The leaves are glossy green and have a pronounced horny edge with prominent brown teeth of up to 10 mm long. The much-branched, robust inflorescence is rather sparsely flowered and can be close to 3 m high. The dull red flowers are typically swollen at the base and, as in the case of many other spotted aloes, have a grey waxy layer on the surface.

The species flowers in February and March.

Aloe dewetii is a very characteristic species, unlikely to be confused with any other spotted aloe. It is a very large plant and does not produce any offshoots (i.e. the plants grow as single individuals). The leaves have a horny edge, pronounced marginal teeth and a glossy sheen.

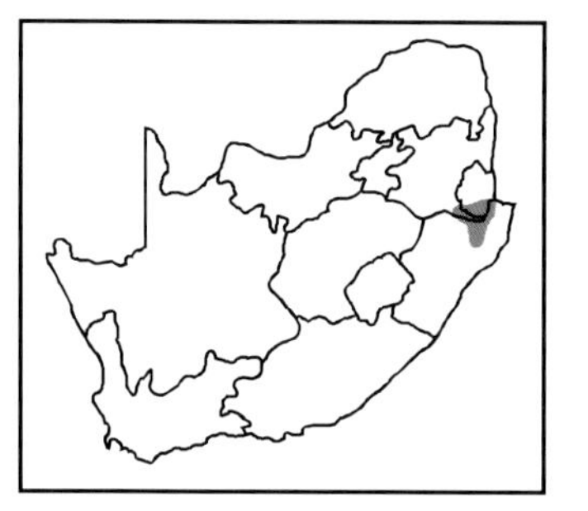

The species occurs in Mpumalanga near Piet Retief, in Swaziland, and in northern KwaZulu-Natal, where it favours open, gently sloping grasslands.

Aloe dewetii is not threatened.

This is a very attractive and robust spotted aloe, but remains uncommon in cultivation. Since it flowers in late summer, the inflorescences are not damaged by frost. Reynolds named the species after J.F. de Wet, who brought it to his attention.

Ben-Erik van Wyk

Ben-Erik van Wyk

Ben-Erik van Wyk

Aloe dyeri

The plant is large for a spotted aloe and generally do not form a stem. They characteristically occur as solitary specimens. The rosettes of mature plants could be up to 2 m in diameter. The deeply channelled leaves are usually distinctly recurved and are dark yellowish-green. The upper and lower leaf surfaces usually have fairly small scattered or transversely arranged whitish spots. The inflorescence is sparsely flowered, branched and up to 2 m high. The flowers are dull red and have the typical basal swelling of spotted aloes.

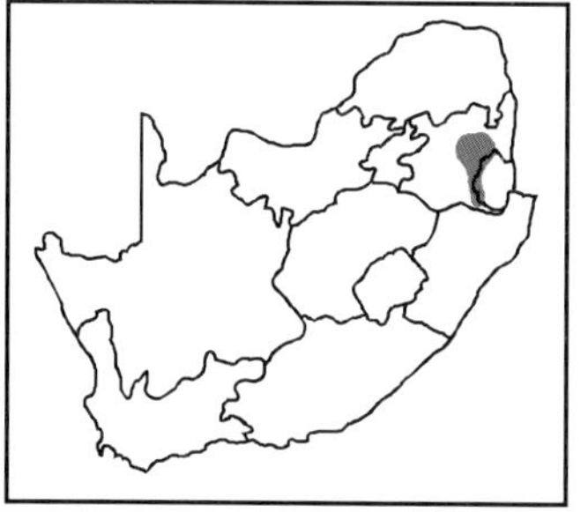

The species flowers from February to June.

Aloe dyeri is very robust and is certainly one of the largest spotted aloes occurring in Mpumalanga.

In the natural habitat, the plants occur in shady areas in river valleys and ravines. They occur in Mpumalanga at high altitudes, from Piet Retief, Badplaas and Waterval Boven in the west to Nelspruit and Barberton in the east.

The species is not threatened.

Aloe dyeri is a shade-loving species. This should be borne in mind if it is to be cultivated successfully. The species was named after Sir William T. Thiselton-Dyer of Kew, and not Dr R. Allen Dyer, well-known South African botanist, who had a special interest in indigenous succulents.

Frits van Oudtshoorn

Frits van Oudtshoorn

Ben-Erik van Wyk

Aloe fosteri

Plants are usually stemless and grow as scattered individuals. The leaves are bluish-green and have a grey, powdery bloom. Like most spotted aloes the leaves die back from the tips, especially in winter. The upper and lower leaf surfaces are indistinctly striped, but only the upper surface has numerous scattered H-shaped spots. The inflorescence is branched from below the middle and bears brightly coloured flowers, ranging from uniformly yellow to orange and even scarlet.

The species flowers in March and April.

The plant is fairly large for a spotted aloe, and can be up to 1 m across. The brightly coloured flowers, reminiscent of miniature candy sticks, are quite distinctive.

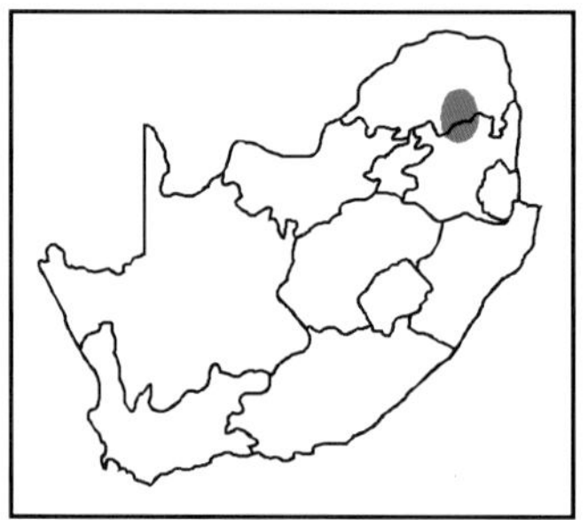

Aloe fosteri occurs in Mpumalanga, to the west of the Drakensberg. Its general distribution range is from Buffelsvlei in the south to near Pietersburg in the north. It generally occurs in open bushveld.

The species is not threatened.

This is arguably the most handsome of all the spotted aloes and makes a beautiful display when planted in groups. It was named after an aloe enthusiast, Cyril Foster.

Pitta Joffe – NBI

Pitta Joffe – NBI

Pitta Joffe – NBI

Aloe grandidentata

Plants are stemless and produce numerous suckers to form large, dense groups. The lance-shaped leaves, arranged in a low, rather flat rosette, are copiously white-spotted, more so on the lower surfaces. Although the spots are arranged in distinct or indistinct bands, they usually do not coalesce. The inflorescence, which is much-branched, can reach a length of about 1 m. The dull red flowers are sparsely dispersed in the racemes. This is the only spotted aloe with club-shaped flowers.

Aloe grandidentata flowers in spring from August to October.

The most important characteristic that distinguishes *A. grandidentata* from all its relatives is the club-shaped flowers. In general appearance, the rosettes resemble those of *A. zebrina*, particularly the form previously recognised as *A. transvaalensis*.

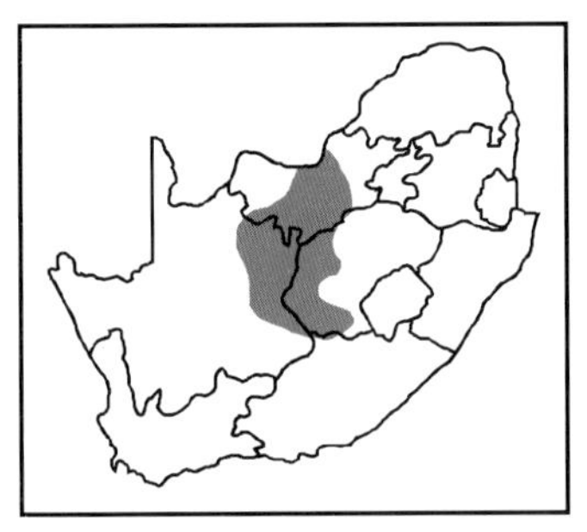

Aloe grandidentata is widely distributed in the arid inland of South Africa. It occurs in the Northern Cape Province around Kimberley, northwards into Botswana, parts of the western and central Free State and the Northwest Province.

The species is common throughout its distribution range and is not threatened.

Aloe grandidentata grows easily and is underestimated as a ground cover. It will quickly cover an open patch in a garden, and will tolerate very low winter temperatures, such as those encountered on the Highveld. Afrikaans common names of *A. grandidentata* are Bontaalwyn and Kanniedood. The scientific name *grandidentata* ('with large teeth') is somewhat misleading, as the thorns of this aloe are not particularly large.

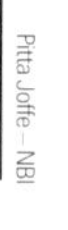

Pitta Joffe – NBI

Ben-Erik van Wyk

Ben-Erik van Wyk

Aloe greatheadii

Plants are stemless, but fairly robust and grow singly or in medium-sized groups of up to fifteen. The firm, fleshy leaves range from triangular to lance-shaped. They are shiny green, often faintly striped and have the oblong white spots arranged in more or less distinct bands. The lower leaf surfaces are usually a dull whitish-green. The leaf margins are armed with sharp dark brown teeth and the leaves tend to die back slightly in winter. The inflorescence can be up to 1,5 m high and has up to six branches. In the case of the var. *davyana*, two inflorescences are commonly formed successively. The typical variety flowers more proliferously and it is not uncommon to find plants producing four inflorescences simultaneously or successively. The flowers range from pale pink to a handsome bright red. The species flowers in midwinter from June to July.

In many parts of its distribution the more commonly found *A. greatheadii* var. *davyana* grows together with *A. zebrina*, particularly the form previously ascribed to *A. transvaalensis*. The leaves of *A. greatheadii* var. *davyana* are, however, lance-shaped, shiny green and carried in a spreading rosette. The rosettes of this species also create the impression of being generally tidier than those of *A. zebrina*. *Aloe greatheadii* var. *davyana* flowers in winter, whereas *A. zebrina* flowers in summer. (Differences between *A. greatheadii* var. *davyana* and the closely related *A. verdoorniae*, which is here doubtfully retained as distinct, are discussed under the latter species.)

Aloe greatheadii var. *greatheadii* has its main centre of distribution in Zimbabwe but also occurs in Botswana, Zambia, Malawi, Mozambique and the Congo. It enters South Africa in the Northern Province where it is found near Louis Trichardt and Soekmekaar. In contrast, *A. greatheadii* var. *davyana* occurs in all the northern provinces of South Africa, including the Free State, where it usually occupies rocky outcrops in grassland. This variety is very common on the Witwatersrand.

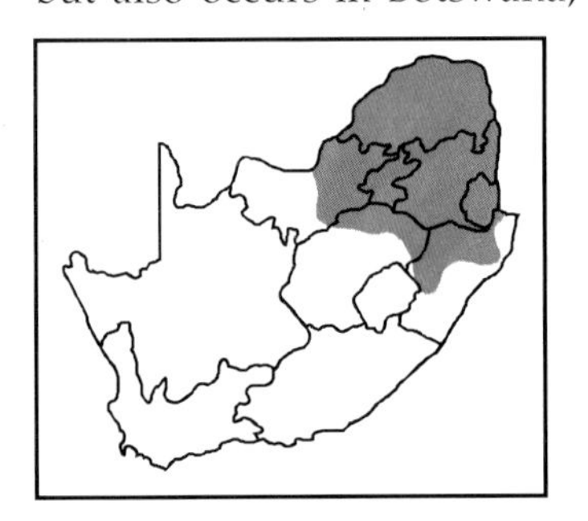

Neither of the varieties are threatened.

The variable var. *davyana* is an important component of the succulent flora of the Witwatersrand. Trials have shown that it can be used successfully as a soil binder in disturbed areas such as mine tailings. Its leaf sap is also used medicinally, particularly for the treatment of burns, sores and wounds. It grows easily from seed and transplants well as mature specimens. If planted in large numbers in full sun or semi-shade, the pink flowers of the variety will brighten up a drab winter Highveld garden. The common name of *A. greatheadii* var. *greatheadii* is Kizima-bupia (Kitabwa, Zaire). For *A. greatheadii* var. *davyana* common names recorded are Kleinaalwyn and Transvaalaalwyn, and Kgopane (seTswana). Spotted aloes that were upheld by G.W. Reynolds in his benchmark publication on *Aloe*, but which are currently included in the variable *A. greatheadii* var. *davyana*, are *A. davyana* var. *subolifera*, *A. barbertoniae*, *A. mutans* and *A. graciliflora*.

Pitta Joffe – NBI

Eben van Wyk

Eben van Wyk

Aloe greenii

Plants are usually stemless and have up to twenty lance-shaped, recurved leaves in a dense rosette. They form thick stands by means of basal suckers. The bright green leaves are up to 500 mm long and are densely white-spotted on both surfaces. The spots are oblong in shape and become confluent in irregular wavy bands. The margins are armed with pinkish-brown prickles. Up to two inflorescences, each about 1,3 m tall, are formed consecutively. The dark pinkish flowers are typically constricted above the ovary and have a basal swelling.

The flowering period is mostly from March to April.

The closest relative of *A. greenii* is *A. pruinosa*. However, the latter species is easily distinguished by the substantial greyish-white, powdery bloom that covers the peduncle and flowers. The single inflorescence of *A. pruinosa* is usually taller than that of *A. greenii*. In the vegetative phase *A. pruinosa* is a larger, solitary plant, and forms clusters only when the growing tip has been damaged.

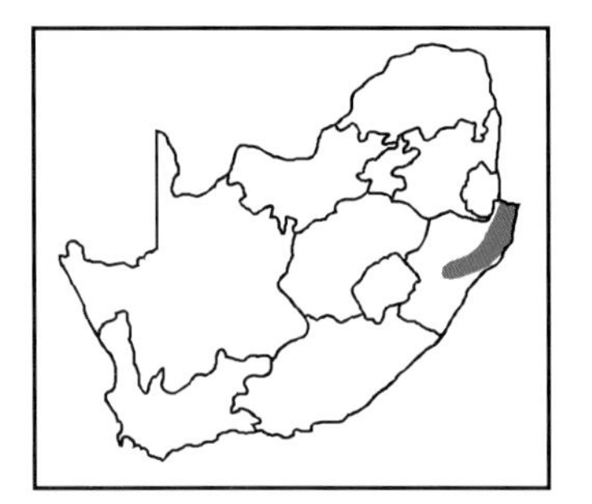

Aloe greenii occurs in central KwaZulu-Natal from near Pietermaritzburg northwards to southern Mozambique. It usually grows in the shade of trees and shrubs.

The species is not threatened.

Aloe greenii is not a very attractive plant in cultivation and will quickly multiply to form an embarrassing number of basal suckers. The Zulu common name of the species is Icena. The origin of the scientific name was never recorded.

Brian Kemble

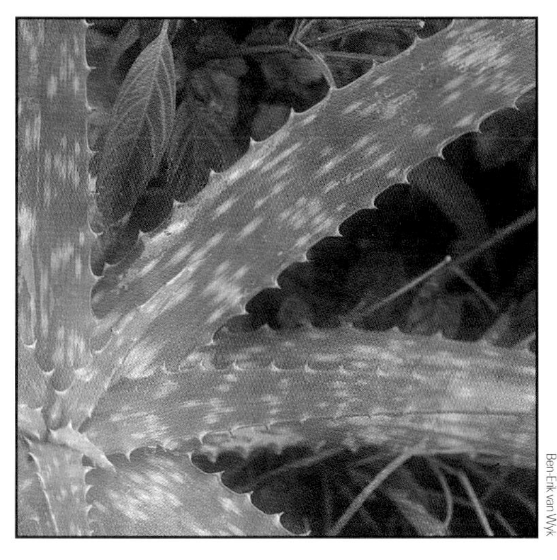

Ben-Erik van Wyk

Brian Kemble

Aloe immaculata

Plants are usually stemless, but may form short, erect stems covered with the remains of old dead leaves. Solitary specimens are commonly found, that is they do not produce any suckers. However, in any population a few suckering plants will be encountered. The dark green to greyish-green leaves usually have spots on the upper surfaces, although plants that do not have any spots are also commonly found. The leaf margins do not have the same horny edges as those of *A. affinis*. The leaves are armed with sharp, reddish teeth. The fairly short inflorescences are much-branched and bear numerous cone-shaped racemes. The flowers are any shade of red, from orangy-pink to coral-red.

Aloe immaculata flowers from June to August.

The species can hardly be distinguished from the other spotted aloes with which it grows. A critical study of the spotted aloes might well prove that it should be grouped with a widespread species such as *A. greatheadii*. It differs from *A. greatheadii* var. *davyana* in being a more robust plant in all respects and the flowers are more brightly coloured.

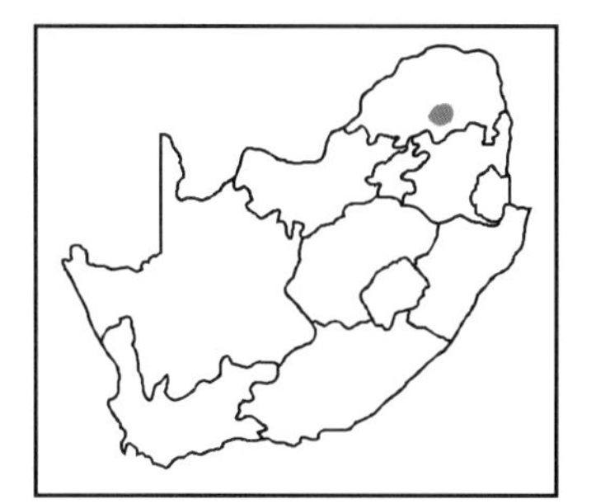

Aloe immaculata occurs from Pietersburg southward to Lydenburg, especially in the warm river valleys. It is particularly common around the little settlement of Chuniespoort. Plants occur in open grassland or in the shade of bushes.

Aloe immaculata is not threatened in its natural habitat.

The species occurs in an area where the distribution ranges of many spotted aloes overlap. It is therefore extremely difficult to decide which species one is dealing with in the distribution range of *A. immaculata*. *Aloe immaculata* is easy to cultivate and can withstand some degree of frost, but it is not a very attractive species and is seldom grown by collectors. The scientific name *immaculata*, which means 'without spots', is rather inappropriate as the leaves are commonly spotted.

Gideon Smith

Gideon Smith

Ben-Erik van Wyk

Aloe lettyae

Plants are solitary and stemless. The rosettes consist of up to twenty erectly spreading, bluish-green leaves. The leaves have dull whitish spots on both surfaces and are armed with small, brown, marginal teeth. The inflorescence could be up to 2 m high and has gracefully curved side-branches. The pinkish-red flowers have distinctly rounded bulbous bases.

Aloe lettyae flowers from February to April.

The species is easily distinguished by its flowers which have large round basal swellings. Furthermore, the leaves are spotted on both surfaces, not just the upper surface as with many spotted aloes. The inflorescence branches have distinctly rounded curves, a characteristic not encountered in any other spotted aloe.

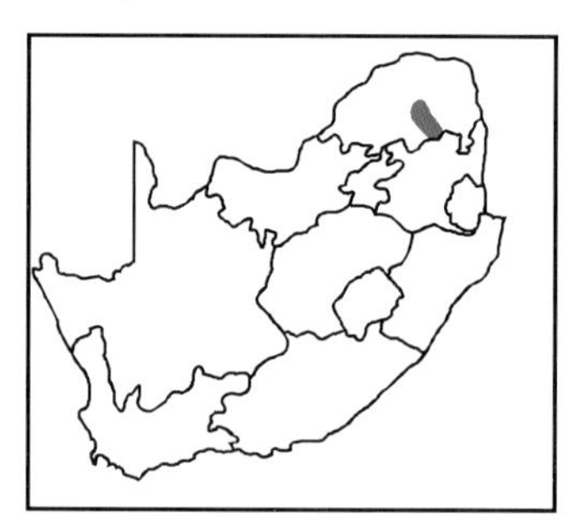

Aloe lettyae occurs amongst tall grass in hilly country in the Northern Province in a fairly restricted area in the vicinity of Duiwelskloof and Magoebaskloof, about 100 km east of Pietersburg.

The species is not threatened.

Aloe lettyae is easy to cultivate and flowers before the early frosts occur. The species was named after Ms Cythna Letty, well-known South African botanical artist, who contributed much to our knowledge of the genus *Aloe*, particularly through her artwork in the *Flowering plants of (South) Africa*.

Ben-Erik van Wyk

Frits van Oudtshoorn

Ben-Erik van Wyk

Aloe longibracteata

This aloe occurs as a small and stemless plant, mostly solitary. The short, closely packed, thick, succulent leaves are more or less triangular in shape and are borne in a basal rosette. The off-white, somewhat H-shaped leaf spots are fairly large and occur on the upper surfaces only. Up to 100 mm of the terminal portions of the leaves are dry and twisted, forming strong thorny tips. The leaf margins, too, are armed with sharp, brown teeth. The inflorescence, which may be up to 1 m high, is branched fairly low down. The racemes are elongated and rather densely flowered. The flowers vary from dull pink to bright red, always with yellowish mouths, and have basal swellings typical of just about all the spotted aloes. The flowers are lightly covered in a whitish, powdery waxy layer and are characteristically subtended by long, curled, white bracts.

The species flowers from July to August.

Like most spotted aloes, *A. longibracteata* is an immensely variable species. It is primarily characterised by having the oldest leaves horizontally disposed and not erect to erectly spreading. Furthermore, the leaves are very thick and triangular in shape. The species differs from other spotted aloes by often having the terminal portions of the racemes slanted to one side, and by the presence of white bracts of up to 50 mm long below each flower.

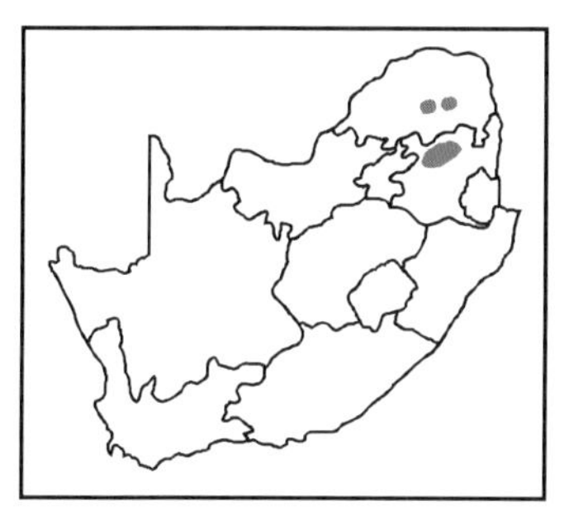

The highveld grassland of Mpumalanga is the prime habitat of *A. longibracteata*. It occurs chiefly in the vicinity of Middelburg, stretching eastwards to Lydenburg and northwards to Pietersburg and Tzaneen.

The species is not threatened.

It has been suggested that the species should be combined with *A. greatheadii* var. *davyana*, but it is provisionally kept here as a separate species. The name *longibracteata* refers to the long floral bracts of the species.

Pitta Joffe – NBI

Eben van Wyk

Eben van Wyk

Aloe maculata

Plants can be stemless or will in time form a short, erect stem of less than 1 m high. The broad, triangular leaves vary considerably in length and shape, but are mostly slightly recurved towards the dried, twisted tips. Leaves are copiously spotted throughout and have sharp, brown, marginal teeth. Leaves are often carried in rather open, flat rosettes. The flat-topped inflorescence can have up to six branches. The individual racemes are distinctly flat-topped, an exception being the form with more rounded racemes from the western Free State, that was previously treated as var. *ficksburgensis*. The stalks of the open flowers are longer than those of the buds. Flower colour ranges from red and orange to yellow.

Flowering time is variable and various forms may flower in summer (January), winter (June) or spring (August, September).

Aloe maculata is fairly distinctive and is usually not confused with other spotted aloes occurring in the same area. Although it is a very variable species, the distinctly flat-topped racemes and usually uniformly coloured flowers distinguish it from most other spotted aloes.

This species has a wide distribution from the Cape Peninsula through the Western and Eastern Cape Provinces, into the eastern Free State and Lesotho, through KwaZulu-Natal and Mpumalanga to the Inyanga District in Zimbabwe. Plants usually prefer the milder coastal climates but are also found as a component of the higher altitude Drakensberg flora. They occur in a variety of habitats, ranging from rocky outcrops to thicket and grassland.

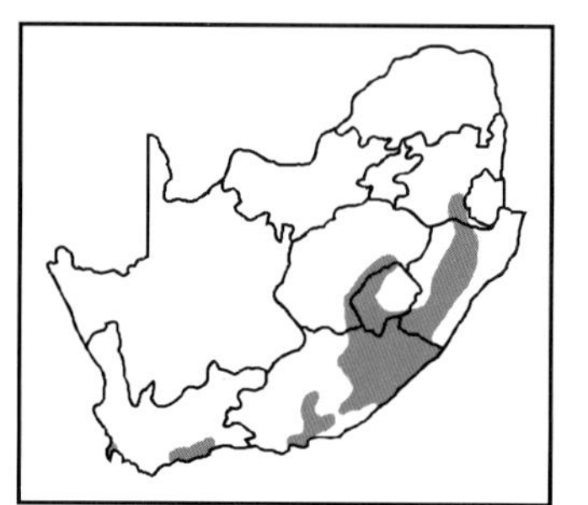

Aloe maculata is common throughout its wide geographical distribution range and is not threatened.

For many years *A. maculata* has been incorrectly called *A. saponaria*. Although the var. *ficksburgensis* from the eastern Free State has been separated from the typical variety, it hardly warrants recognition in a taxonomic hierarchy. *Aloe maculata* is very common in cultivation throughout South Africa. Its wide distribution range indicates that it can tolerate a variety of soil types and watering regimes. It hybridises readily with a number of other aloes, both in habitat and in gardens. A cross with *A. striata* is particularly common. This rather attractive hybrid with its finely toothed, pink leaf edges is probably the most widely cultivated aloe hybrid and should be given formal recognition as a cultivar. Common names of *A. maculata* include Common soap aloe and Bontaalwyn, and Icena (Zulu). *Maculata* means 'spotted' or 'blotched'.

Ted Scholes

Dave Hardy

JLC Marais

Frits van Oudtshoorn

Aloe monotropa

Plants have short, creeping stems of up to 300 mm long. Suckers are sometimes produced along the length of the stem. The leaves have irregular white spots and also lines on them. In contrast to many other spotted aloes, the leaves are more or less erect in a loose rosette, with the terminal portions of the leaves variously recurved. Old leaves are persistent. The inflorescence does not grow very tall (about 800 mm), and more or less horizontal branches arise from about the middle or higher up. The flowers are reddish pink or occasionally yellow, and are characteristically borne vertically.

Flowering is from September to December.

The species is immediately recognisable by its vertical flowers, which all point in the same direction. When not in flower, *A. monotropa* can be distinguished by its striped, almost snake-like leaves.

Aloe monotropa has a restricted distribution in the Dublin Mine Kloof in the Northern Province. It occurs in a well-wooded area and is typically a species of the forest fringe where it occurs on steep rocky slopes.

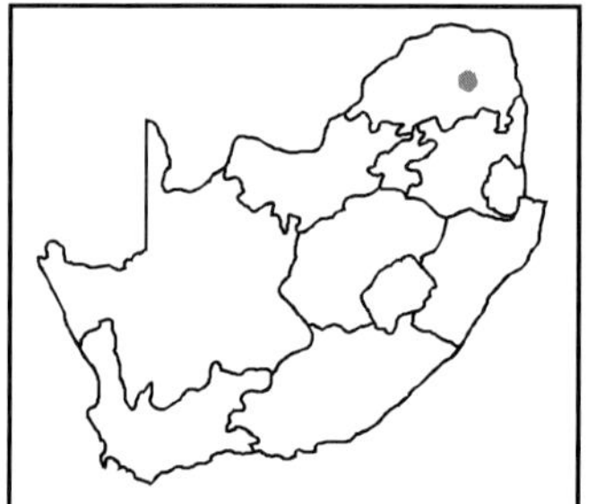

Aloe monotropa is regarded as endangered, due to its small world population and pressure by illegitimate collection.

This is a very distinctive species, even when not in flower, hence the scientific name which means 'alone and on its own'. The late discovery (in 1958) is certainly due to the very restricted geographical distribution range. *Aloe monotropa* is easy to cultivate and will flower even if it is grown very far from its natural habitat.

Steve Fourie

Peter Raal

Ben-Erik van Wyk

Brian Kemble

Aloe mudenensis

Plants are usually solitary or may form small clumps of up to ten individuals. Spotted aloes are usually more or less stemless, but in this species the plants will commonly form a stem of about 800 mm high. The leaves are a bluish-green colour and have numerous scattered spots, particularly on the upper surface. The inflorescence is branched, with up to eight racemes. Flower colour varies greatly from yellowish-orange to red, with a light pinkish-orange being most commonly encountered.

This is a fairly distinctive spotted aloe, which is particularly characterised by its cylindrical, yet terminally rounded racemes. Additional distinguishing characteristics are the short erect stems, the dry leaves that do not remain on the plant and a cut leaf surface which tends to turn purplish when dry.

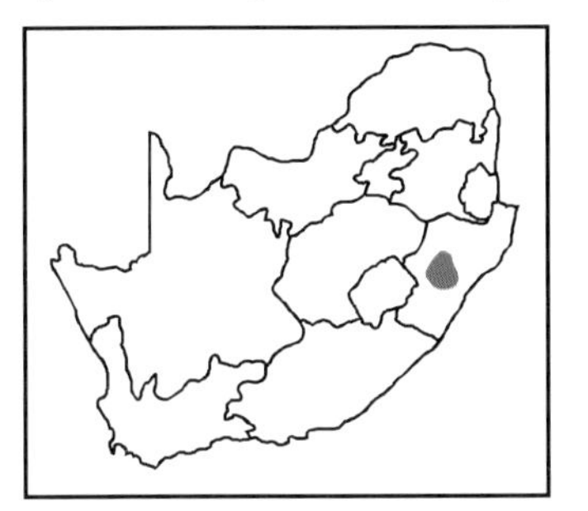

The species flowers in July.

Aloe mudenensis occurs in northern KwaZulu-Natal, particularly around the town of Muden, after which it was named.

The species is not threatened.

Aloe mudenensis is one of the more attractive species of spotted aloes. It is fairly easy to cultivate, but should be protected from severe frost.

Ben-Erik van Wyk

Ben-Erik van Wyk

Ben-Erik van Wyk

Aloe parvibracteata

Plants are generally stemless, but may in time develop a short, erect or creeping stem. Leaf arrangement and shape are variable and range from comparatively short, erect and triangular to fairly long, twisted and sickle-shaped. Leaf colour varies from bright green to distinctly purplish-green. The upper leaf surfaces have numerous dull white spots. The leaf margins are armed with well-spaced triangular prickles. Up to three sparsely flowered inflorescences are produced consecutively. Flowers can be any shade from bright orange to bright red. Plants with orange or red flowers often occur side by side.

Aloe parvibracteata flowers in June and July.

In habitat *A. parvibracteata* looks like a typical spotted aloe and it has few distinguishing characteristics. The form of the species which is most widely cultivated is easy to distinguish from all other cultivated spotted aloes by its dark purplish-green leaves and orange flowers.

Aloe parvibracteata grows on rocky outcrops in grassland and amongst trees and bushes in warm, usually low-lying areas and river valleys in the southern part of Mpumalanga around Komatipoort. It also occurs in Mozambique (this form was previously called *A. lusitanica*), Swaziland and KwaZulu-Natal which appears to be the centre of the distribution area.

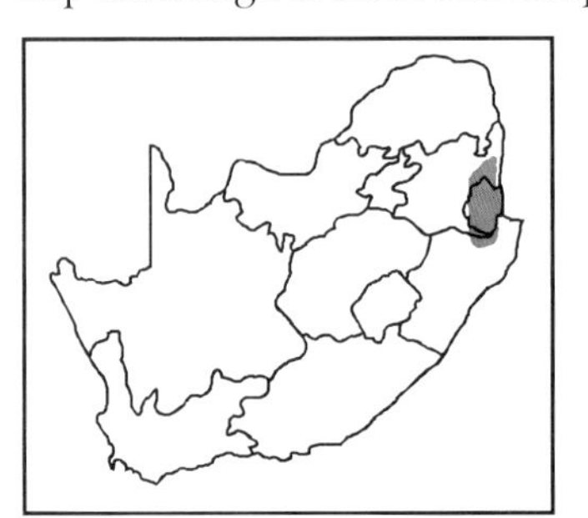

Aloe parvibracteata is not threatened anywhere in its distribution range.

The species is very easy to cultivate and immensely proliferous from the base. The form with purplish-green leaves and waxy orange flowers is a common garden subject, particularly in Highveld gardens. It is not a very tidy-looking plant and is not sought after by collectors. In our opinion, the var. *zuluensis* does not warrant separate recognition and it is here included in this variable species, along with *A. komatiensis*. The scientific name *parvibracteata*, which means 'small bracts', is somewhat misleading since many plants do not have particularly small bracts below the flowers.

Frits van Oudtshoorn

Pitta Joffe – NBI

Pitta Joffe – NBI

Pitta Joffe – NBI

Aloe petrophila

These stemless plants are small in comparison with most spotted aloes, and grow singly or in small clumps. The bright greenish-brown, recurved leaves are arranged in dense rosettes and have reddish teeth along the margins. The inflorescence is branched, with up to six more or less head-shaped, loosely flowered racemes. The flowers are an attractive bright pink.

The flowering time is from May to June.

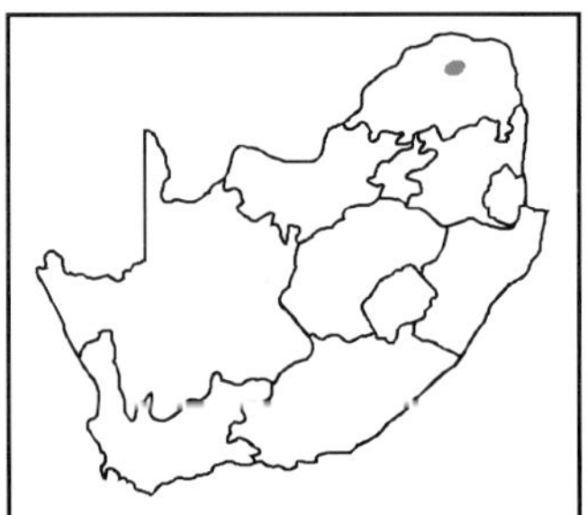

This is one of the smallest of all the spotted aloes and can easily be identified when the rocky habitat and geographical origin are known.

Aloe petrophila has a restricted distribution in the Soutpansberg in the Northern Province, where it often grows on sheer rock faces.

Aloe petrophila is not common in habitat and due to its localised distribution should be listed as rare.

The species is fairly easy to cultivate, provided that it receives adequate protection from severe frost. The scientific name *petrophila* means 'lover of rocky places'.

Pitta Joffe – NBI

Leo Thamm

Pitta Joffe – NBI

Aloe prinslooi

Plants are stemless and solitary although some specimens will form small clumps of several rosettes from a single rootstock. The very succulent leaves have variable shapes, but they are most often distinctly triangular. The leaves have white, oblong spots on both surfaces. These are denser on the upper surfaces. The inflorescence is branched, with up to four dense, head-shaped racemes. The flowers are rather small and do not have the characteristic basal swelling which is generally associated with spotted aloes. Flower colour ranges from creamy to pinkish-white.

Flowering is from June to October.

Plants of this species are easy to recognise when in flower. The flowers are the least like those associated with spotted aloes, in that the prominent bulbous base is lacking. Furthermore, the flowers are creamish to pinkish-white and superficially resemble those of some of the small grass aloes, such as *A. minima*. The

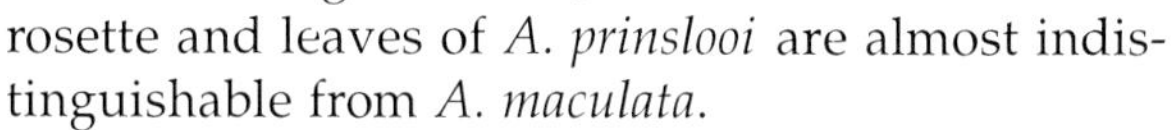

rosette and leaves of *A. prinslooi* are almost indistinguishable from *A. maculata*.

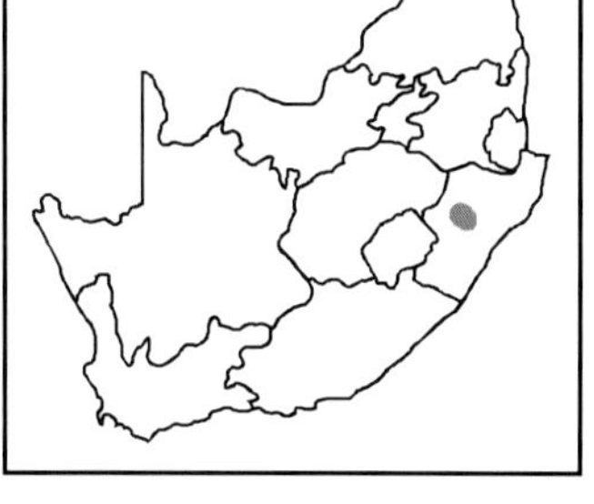

Aloe prinslooi grows on grassy slopes in parts of central KwaZulu-Natal, especially around Muden, Colenso and Ladismith.

The species is regarded as rare mainly due to injudicious collecting by succulent enthusiasts.

Aloe prinslooi was discovered by G.J. Prinsloo, some years after the publication of Reynolds's book on South African aloes. In most respects it resembles various other spotted aloes and it is quite understandable why it went undetected for so long. The species is one of the more difficult spotted aloes to cultivate successfully and is best left in its natural habitat.

Pitta Joffe – NBI

Ben-Erik van Wyk

Pitta Joffe – NBI

Aloe pruinosa

Plants usually have a short, creeping or semi-erect stem of up to 500 mm long, with about thirty recurved leaves in a dense rosette. Individuals of the species occur as solitary plants. The bright green leaves are up to 700 mm long and are densely white-spotted on both surfaces. The spots are rather H-shaped and become confluent in irregular wavy bands, particularly on the lower sides of the leaves. The margins have pinkish-brown prickles. A single, much-branched inflorescence, about 2 m high, is formed. The dark pinkish flowers are typically constricted in the middle and have a basal swelling. The flowers and peduncle are covered with a conspicuous greyish-white powdery bloom.

The flowering period is from February to March.

Of all the species of *Aloe*, the inflorescence and flowers of *A. pruinosa* are the most thickly covered with a white powdery layer of wax. An additional interesting diagnostic characteristic is that the leaf sap of this species rapidly dries to a deep purple. The closest relative is *A. greenii*. (See the latter for a discussion of the differences between the species.)

The species is restricted to the Pietermaritzburg district in central KwaZulu-Natal, where it occurs in the shade of trees and shrubs.

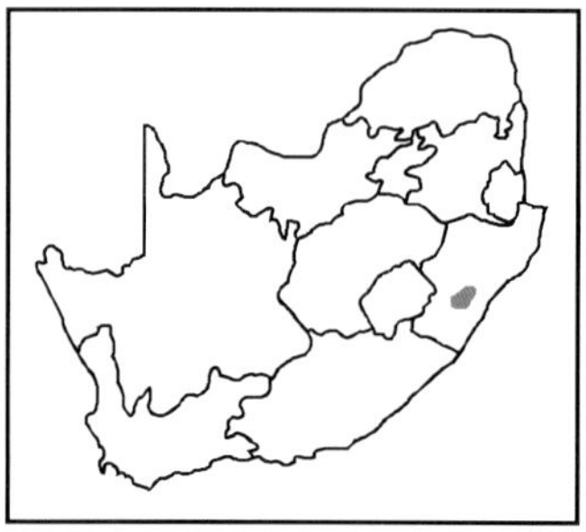

Aloe pruinosa has a more localised distribution than *A. greenii*, and unlike the latter, its conservation status is rare. Although it occurs in the Bisley Valley Nature Reserve south of Pietermaritzburg, its continued survival is not guaranteed as urban expansion and industrial development are rapidly encroaching even on this locality.

The species is slightly more difficult and grows more slowly in cultivation than *A. greenii*. It also does not form basal suckers and will therefore not colonise a denuded area or garden bed as rapidly as *A. greenii*. The Zulu common name for *A. pruinosa* is Icena elikhulu ('the big icena'). In Afrikaans it is known as Kleinaalwyn or Slangkop. *Pruinosa* means 'covered in a waxy white powdery bloom'.

Neil Crouch

Gideon Smith

Neil Crouch

Aloe simii

Plants are stemless and solitary, that is they do not form suckers or clumps. The bright green to milky green leaves are up to 600 mm long and 120 mm wide at the base and are distinctly channelled. The leaves are most often unspotted, although a few faint scattered spots are sometimes found, particularly on the upper surface. The lower leaf surfaces usually have obscure longitudinal lines. The inflorescence can reach a length of between 1,5 and 2 m and is branched and rebranched, with up to fifteen sparse racemes. The flowers have large basal swellings and are pink.

The flowering period is from January to March.

Aloe simii is characterised by having more or less unspotted, robust and deeply channelled leaves that are usually more or less erect. For this reason the species is sometimes confused with a small species of exotic *Agave* (Century plant). The sparsely flowered racemes and inflorescences with widely spreading branches are also typical of *A. simii*. The species has a restricted distribution in Mpumalanga and usually occurs in tall, open grassveld. It used to occur abundantly around White River, extending to the Sabie River Valley near Sabie and southwards to Rosehaugh and Nelspruit, but afforestation with exotic trees has unfortunately led to a decrease in suitable habitats.

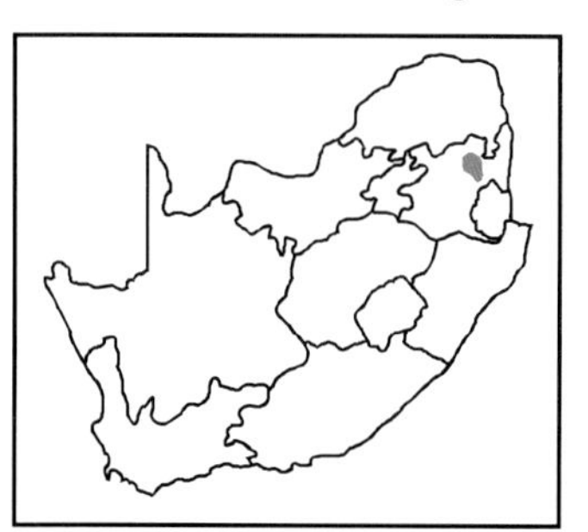

Aloe simii is at present not considered to be threatened.

Due to its characteristic leaf morphology, this is one of the more distinctive species of spotted aloes. Like *A. dewetii* and *A. dyeri*, *A. simii* is a large plant with a very tall inflorescence. It is fairly easy to cultivate, provided that the plants are protected against frost and are watered adequately. *Aloe simii* was named after T.R. Sim, who first collected the species.

Frits van Oudtshoorn

Frits van Oudtshoorn

Frits van Oudtshoorn

Aloe suffulta

Plants are solitary and have short stems of up to 200 mm long. The stems carry leaves from ground level upwards and have internodes of up to 10 mm long. The gracefully recurved leaves are dark green or brownish green, with dull white spots on both surfaces. The upper surfaces are deeply channelled and the leaf margins have white, harmless teeth. The inflorescence is weak, sparsely flowered, branched and up to 2 m high. Each rosette produces only one inflorescence. The flowers are a dull pinkish-red.

The species flowers in June and July.

In contrast to all the other spotted aloes, *A. suffulta* has a slender twining inflorescence. This unusual characteristic separates it from all other aloes.

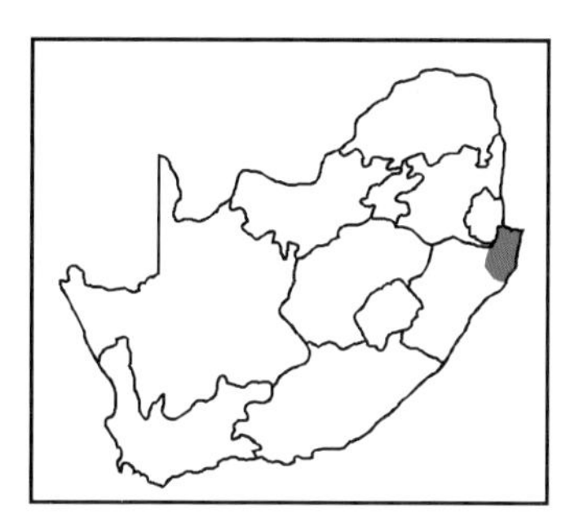

Aloe suffulta thrives in sandy soils in shady positions amongst bushes in northern KwaZulu-Natal and Mozambique.

The species is not threatened.

Aloe suffulta is uncommon in collections of succulent plants and does not thrive in cultivation away from its natural habitat. This species is aptly named, because *suffulta* means 'supported' or 'propped', referring to the weak and slender inflorescences which are always supported by the surrounding trees.

Pitta Joffe – NBI

Leo Thamm

Ben-Erik van Wyk

Aloe swynnertonii

Plants are stemless and grow singly or in small clumps. Depending on where the plants grow, the leaves vary from 250 mm to about 1 m in length. The upper leaf surfaces are dark green and copiously white-spotted. The lower leaf surfaces are usually unspotted and lighter green in colour. Leaf margins carry well-spaced reddish teeth. The inflorescence has up to fifteen branches and is usually between 1,5 and 2 m high, but can reach a height of 2,5 m in robust specimens. Individual racemes are distinctly flat-topped and densely flowered. Flowers are usually red, but plants with orange flowers are sometimes encountered. The bases of the flowers are markedly swollen.

Flowering occurs from April to July.

Aloe swynnertonii is easily recognisable due to its narrow, lance-shaped leaves and tall, head-shaped inflorescences. Although not a consistent characteristic, the lateral racemes of the species are often taller than the terminal ones.

The species is not as widely distributed in South Africa as most other spotted aloes, because the centre of the distribution area lies north of South Africa. It occurs in the Northern Province, eastern Zimbabwe, Mozambique and Malawi. Plants are found in widely differing habitats, ranging from the fringes of forests, to grassy and rocky slopes.

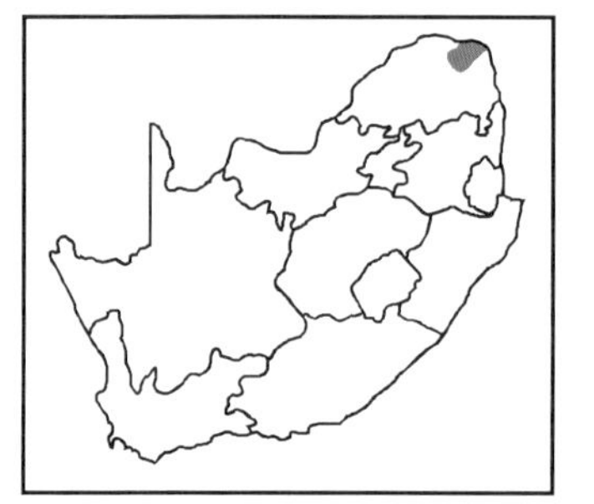

Aloe swynnertonii is not threatened in its natural habitat.

This robust spotted aloe is easy to cultivate and the graceful, long leaves and tall, dense racemes will make it a useful addition to any garden. It should, however, be noted that the species favours areas with a fairly high rainfall and warm winters. Protection against frost is therefore a prerequisite for the successful cultivation of *A. swynnertonii*. Northern forms of the species were previously called *A. chimanimaniensis*, but this name is no longer used because the plants clearly belong to *A. swynnertonii*. The first specimens were collected in Zimbabwe by C.F.M. Swynnerton, after whom the species was named.

Pitta Joffe – NBI

Aloe umfoloziensis

This spotted aloe is usually stemless or may have short stems of about 400 mm long that are often hidden by the surrounding vegetation. Plants sucker freely. The leaves are fairly long, up to 300 mm, and have a dark, shiny green to brownish-green colour. The spots are dull white or light green and often less conspicuous than in most spotted aloes. Both leaf surfaces may have spots, but the upper surfaces are always more distinctly spotted than the lower. The horny marginal spines are sharp and vary from a greenish to a brownish-green colour. The inflorescence is robust and up to 1,5 m high. The peduncle branches from the middle or higher up into between five and eight racemes, which are more or less head-shaped. The flowers are bright red.

The species flowers from July to August.

This species is characterised by its tall inflorescences that bear round-topped racemes. These are usually smaller than those of *A. maculata*, a species that has been likened to *A. umfoloziensis*. At the northern limits of its distribution range, *A. umfoloziensis* grows with *A. parvibracteata*, which generally has longer, more cylindrical racemes.

Aloe umfoloziensis occurs from near Melmoth and Eshowe in KwaZulu-Natal northwards through the Pongola River Valley to near Mugut and Abercorn Drift. Its prime habitat is the savannah-like vegetation (scattered trees and grass) of the river valleys in north-eastern KwaZulu-Natal.

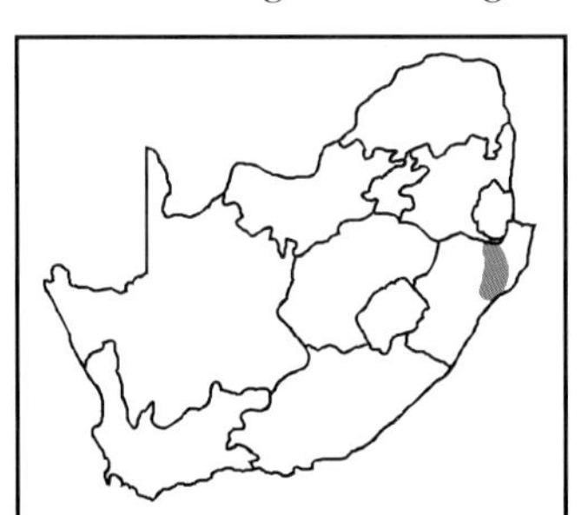

The species is not threatened.

It has been suggested that *A. umfoloziensis* is only a robust KwaZulu-Natal form of *A. maculata* and that it should consequently be included with the latter. However, it is more distinct than some of the other spotted aloes and is therefore provisionally retained. The name of the species was taken from the Black and White Umfolozi Rivers which appear to be the main centre of distribution.

Ben-Erik van Wyk

Ben-Erik van Wyk

Ben-Erik van Wyk

Aloe verdoorniae

Plants are small, solitary and stemless. The leaves are slender and have a peculiar greyish-blue colour and pronounced reddish margins. The leaves are typically obscurely spotted. Up to two sparingly branched inflorescences, usually less than 1 m high, are formed. Inflorescence branching occurs low down. The flowers vary from dull pinkish to red and have a powdery waxy layer.

Aloe verdoorniae flowers in June and July.

The species is closely related to *A. greatheadii* var. *davyana*. It can be distinguished from the latter by its more slender, bluish leaves with contrasting reddish margins. It has been suggested that *A. verdoorniae* should be combined with *A. greatheadii* var. *davyana* and it is therefore doubtfully retained here as a distinct species.

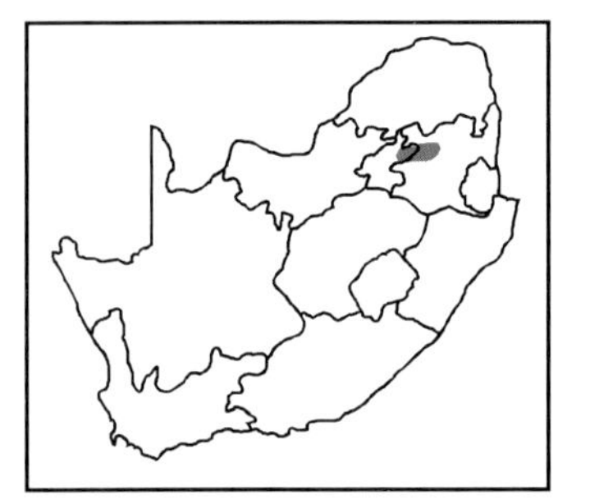

Aloe verdoorniae occurs in grassland from near Cullinan in the west, eastward to Witbank, Dullstroom and Belfast.

The species is not threatened.

This is a typical Highveld species of spotted aloe. The leaf tips are twisted and die back for up to 100 mm. It is generally more attractive than the more common *A. greatheadii* var. *davyana* and *A. zebrina*. The scientific name commemorates the well-known South African botanist, Inez Verdoorn.

Pitta Joffe - NBI

Gideon Smith

Pitta Joffe - NBI

Aloe vogtsii

Plants are usually solitary, but groups of up to eight rosettes formed by basal offshoots are sometimes found. Mature specimens have stems of up to 200 mm high. The pale green leaves are erectly spreading and copiously spotted with numerous minute, white H-shaped spots on the upper surfaces and dark green specks on the lower surfaces. The inflorescence is branched into several racemes. The flowers are bright red and are borne almost horizontally.

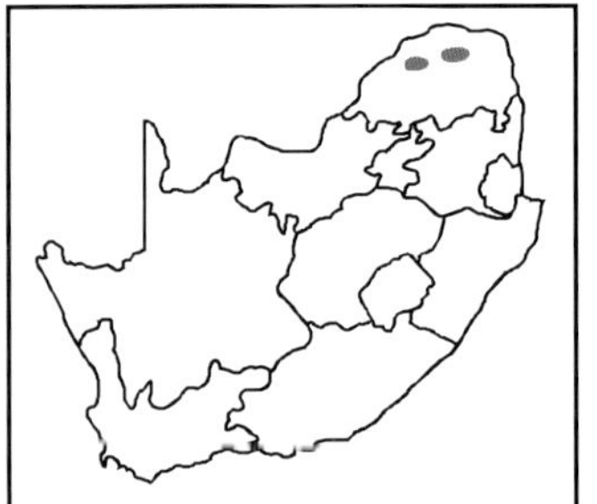

The species flowers from February to April.

This is a very distinctive spotted aloe, easily recognised by the flowers which are borne at an angle of almost 90° to the vertical peduncle. The copiously spotted leaves (both surfaces) also serve to distinguish it from all other spotted aloes.

Aloe vogtsii occurs in tall grass or amongst shrubs and bushes along the top of the Soutpansberg in the Northern Province.

This species is not threatened.

Aloe vogtsii occurs in the mist belt and is restricted to the Soutpansberg. The species was named after its discoverer, Louis R. Vogts.

Richard Watmough

Ernst van Jaarsveld – NBI

Aloe zebrina

Plants are usually proliferous and form dense groups, although solitary specimens are occasionally encountered. The leaves are borne in a small, dense rosette, and vary from lance-shaped to triangular. Leaf colour varies greatly, but tends to be a more dull greyish-green than in associated spotted species. The inflorescences are much-branched and carry the dull pink flowers in rather sparse racemes.

The species either flowers in midsummer (November, December) or late summer (January to March).

Due to the immense variability of the species in terms of size, leaf spots, inflorescences and flowers, it is difficult to give distinguishing characteristics. (Differences between *A. zebrina* and *A. greatheadii* var. *davyana*, both of which occur on the Witwatersrand, are listed under the latter variety.)

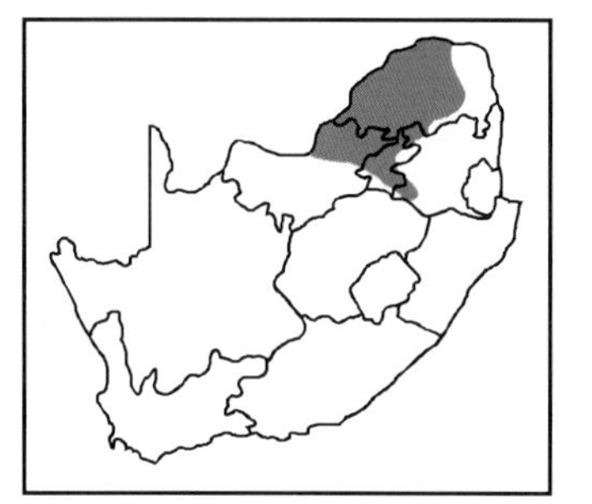

Aloe zebrina is widely distributed in the northern parts of South Africa. It also has an extensive distribution range in Botswana, Namibia, Angola, Zimbabwe and Mozambique.

The species is common throughout its distribution range and is not threatened.

The variability of *A. zebrina* has led to a number of species names reflecting the different manifestations of characteristics. These include *A. ammophila* from around Pietersburg in the Northern Transvaal, *A. transvaalensis* from Pretoria in Gauteng and *A. vandermerwei* from between Leydsdorp and Gravelotte in Mpumalanga. All these are currently included in *A. zebrina*. The scientific name of the species refers to the spots on the leaves, which often merge to form more or less regular stripes or bands. *Aloe zebrina* is not a very attractive species and is seldom cultivated.

Frits van Oudtshoorn

Frits van Oudtshoorn

Pitta Joffe – NBI

*D*WARF ALOES

Group 9

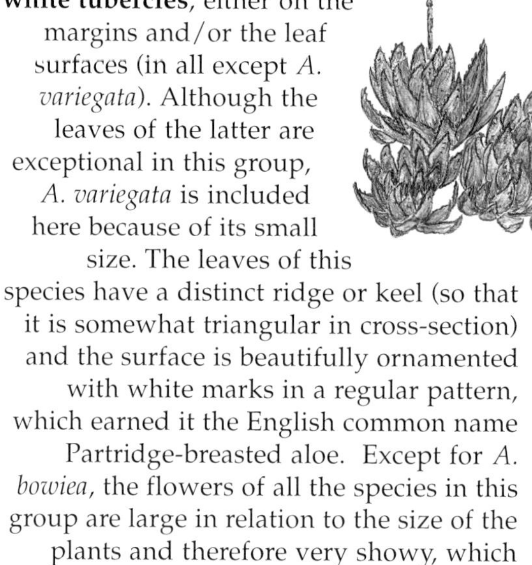

This group of miniature aloes is characterised by the exceptionally **small rosettes** that are usually arranged in groups, and are rarely single except in young plants. The **narrow**, often **incurved leaves** have small, **raised white tubercles**, either on the margins and/or the leaf surfaces (in all except *A. variegata*). Although the leaves of the latter are exceptional in this group, *A. variegata* is included here because of its small size. The leaves of this species have a distinct ridge or keel (so that it is somewhat triangular in cross-section) and the surface is beautifully ornamented with white marks in a regular pattern, which earned it the English common name Partridge-breasted aloe. Except for *A. bowiea*, the flowers of all the species in this group are large in relation to the size of the plants and therefore very showy, which make them highly sought-after collector's items.

SPECIES

A. aristata
A. bowiea
A. brevifolia
A. humilis
A. longistyla
A. variegata

Ernst van Jaarsveld – NBI

Aloe brevifolia

Aloe aristata

Plants are stemless and usually occur in dense groups of up to twelve rosettes. Each rosette is 100 to 150 mm in diameter and has a large number of narrow triangular leaves, which taper into long dry awns. The upper and lower leaf surfaces are dark green with scattered small white spots, some of which are tuberculate with small soft spines, particularly towards the terminal part of the leaf. The leaf margins have rows of soft white teeth, 1 to 2 mm long. Inflorescences are up to 700 mm high, rarely simple, usually divided above the middle into two to six branches, each with a sparse raceme of twenty to thirty flowers. The dull red or pinkish flowers are tubular and slightly curved downwards, with a basal swelling.

Flowering occurs in November.

Aloe aristata is easily recognised by the numerous, white-spotted leaves, the long thin needle-like leaf tips and the tubular, fused petals forming a basal swelling around the ovary. There are no papery bracts along the peduncle as in most other dwarf aloes. When not in flower, the plant is often mistaken for a species of *Haworthia*.

The natural distribution is remarkably wide, both geographically and ecologically. It occurs from the central and eastern Karoo eastwards to the Eastern Cape Province, the Free State, Lesotho and KwaZulu-Natal.

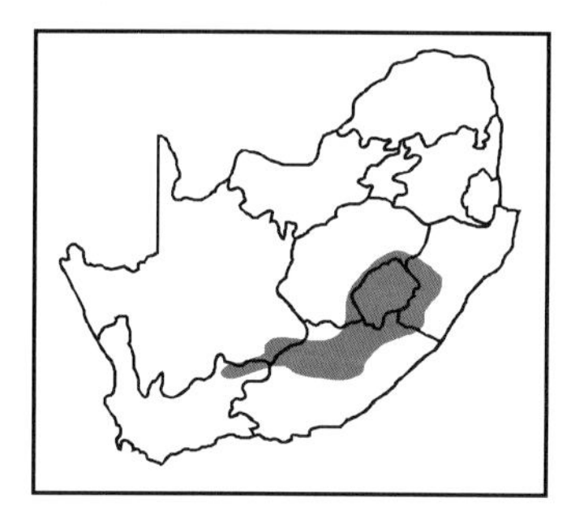

This species is not threatened, but has become extinct at some localities.

In its natural habitat, *A. aristata* is not very attractive because the leaves are usually shrunken and curved inwards. In cultivation, however, the leaves become swollen and spreading, and the plant is popular for rock gardens and grows well in containers. Common names that have been recorded for *A. aristata* include Serelei, which means 'the slippery one'. The scientific name *aristata* means 'awned', referring to the awn-like leaf tips.

Forsyth van Nierop

Geoff Nichols

Neil Crouch

Aloe bowiea

Plants are very small, reaching a maximum height of 140 mm. Solitary specimens are rarely encountered. Several individuals may form dense mats of up to 500 mm in diameter. The leaves are broad at the base, but very thin and wiry upwards. Leaf colour is a uniform pale green and leaves have minute whitish spots near the bases. The leaf margins have minute teeth. An unbranched raceme of about 250 mm tall is produced. The flowers are small and dull greenish-brown.

Especially in cultivation, *A. bowiea* has a tendency to flower throughout the year, but the summer months (November, December and January) are the peak flowering time.

This miniature aloe is very easy to recognise. It is in all respects unlike any other aloe. The roots are typically swollen in the middle, tapering to both sides. The leaves are very thin above ground, but broaden considerably towards the base, giving the plant a bulb-like appearance. The thin peduncle and small, greenish-brown flowers further serve to distinguish the species from the other miniature aloes.

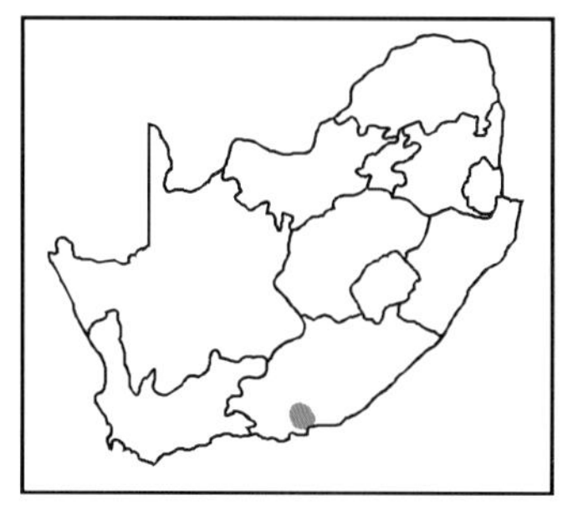

Aloe bowiea is restricted to a small area in the Eastern Cape Province around Port Elizabeth and Kariega. It is a component of karroid patches in valley bushveld vegetation. In pristine condition this vegetation type forms very thorny, impenetrable thickets.

Aloe bowiea is endangered in its natural habitat due to urban and industrial expansion, and overcollection.

In the past, this very distinctive miniature aloe was called *Chamaealoe africana*, but it is not all that different from other aloes and should rather be included in the genus *Aloe*. The species was named after James Bowie, an early plant collector. It is easy to cultivate, but is not a very attractive species.

Ben-Erik van Wyk

Frits van Oudtshoorn

Ben-Erik van Wyk

Aloe brevifolia

Plants usually form large groups of densely leaved rosettes, each about 80 mm in diameter, but up to 300 mm in one variety (see below). Leaves are broad and triangular, about 60 mm long and 20 mm wide at the base and distinctly greyish-green in colour. The upper surface is usually without any spots or spines, but the lower surface may have a few soft spines along the upper middle or scattered on the terminal part. Along the edges of the leaves triangular white spines of up to 3 mm long are found, more or less regularly arranged, approximately 10 mm apart. One or two unbranched inflorescences are formed, usually 400 mm but up to 600 mm high. As with several other dwarf aloes, large bracts are present along the peduncle, almost to the base. The racemes are cone-shaped and rather sparse lower down, but with the buds densely packed and hidden by the bracts. The flowers are various shades of red or sometimes yellow. They are up to 40 mm long, tubular in shape but slightly curved, with the stamens protruding slightly from the mouth of the tube.

Flowering occurs mainly in October and November.

Aloe brevifolia may be distinguished from other dwarf aloes by the combination of the clustered growth form and the relatively broad and thick leaves.

The species is restricted to the Western Cape Province, and occurs naturally in the districts of Caledon and Bredasdorp, extending eastwards to Swellendam, Riversdale and Cape Agulhas. It is found in areas with a relatively high rainfall (about 400 mm per year) and usually grows on clay soil in rocky places.

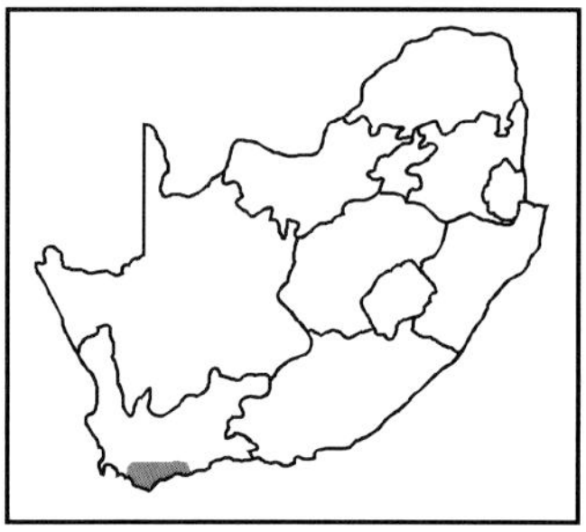

Aloe brevifolia is not threatened but some of the localised forms of the species may become extinct as a result of agricultural development and overexploitation by plant collectors.

Three varieties, based mainly on differences in size, have been recognised. The largest of the three is the var. *depressa*, with rosettes of up to 300 mm in diameter and leaves of about 150 mm long and 60 mm wide. The var. *postgenita* is intermediate in size between this and the typical variety. *Aloe brevifolia* is commonly cultivated and thrives in well-drained soil. The scientific name was well chosen: *brevifolia* means 'short leaves'. The common names are Duine-aalwyn or Kleinaalwyn.

Ben-Erik van Wyk

Ben-Erik van Wyk

Ben-Erik van Wyk

Aloe humilis

This small aloe is a stemless plant and usually occurs in dense groups of up to ten or more rosettes. The leaves are more or less triangular, erect or slightly incurved, with a distinct waxy layer on the surface. They are about 100 mm long and 12 to 18 mm wide. Irregularly arranged white tubercles with soft white prickles occur on the upper surface and more numerously on the lower surface. Soft white teeth of 2 to 3 mm long are present along the margin of the leaf. The inflorescence is always single, up to 350 mm high, with a sparse, oblong raceme of about twenty flowers. The flowers are rather large for such a small plant, being about 40 mm long. They are scarlet or sometimes orange or yellow, tubular in shape but narrowing slightly to the mouth and also to the base. The stamens are situated more or less in the mouth of the flower and do not protrude markedly.

Flowering occurs in August and September.

Aloe humilis is similar to *A. longistyla* in general appearance and may easily be confused with it when not in flower, as both species have a similar geographical distribution. *Aloe humilis* has smaller rosettes which form dense clusters of up to 400 mm wide and which are not predominantly single as in *A. longistyla*. *Aloe brevifolia* also forms dense clusters, but can be distinguished from *A. humilis* by the shorter and broader leaves which do not have prickles or spines on the upper surfaces.

The natural distribution corresponds with dry areas and is reported to be from Mossel Bay in the west, through the Little Karoo to Grahamstown in the east, and northwards to Somerset East and Graaff-Reinet.

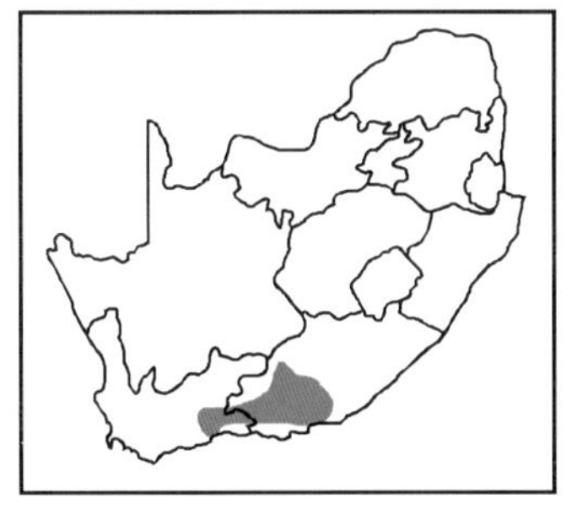

The species is widely distributed and is not threatened, but overcollection and urban expansion may cause serious damage at some localities.

Aloe humilis is rather variable, and several varieties have been recognised. Most of the variation seems to be related to the conditions under which the plants grow. In flat open areas the leaves are usually distinctly grey and incurved. On rocky slopes and in the shade of bushes the leaves are usually larger and erect or slightly spreading, with a greener colour. Unlike *A. longistyla*, this species grows and flowers well in cultivation. The small size of the plant is reflected in the scientific name *humilis*, which means 'low growing'.

Ben-Erik van Wyk

Aloe longistyla

This small aloe is a distinctive stemless plant, usually solitary or occasionally with two to three rosettes, rarely in clusters of up to ten. Leaves are densely crowded, greyish-green, with a distinct waxy layer on the surface. They are up to 150 mm long and 30 mm wide near the base. Both surfaces have firm white spines of up to 4 mm long, each of which has a white tuberculate base. Rows of spines are also present along the margins and sometimes along the middle of the lower surface, but then only towards the terminal part of the leaf. The inflorescences are single or in pairs, up to 200 mm high, with very thick unbranched peduncles and dense, broad cone-shaped racemes of up to fifty flowers. Relative to the small size of the plant, the pale salmon-pink to coral-red flowers are exceptionally large, being up to 55 mm long and 10 mm in diameter. The upper half of the flowers is characteristically curved upwards and the unusually long stamens and style protrude for more than 20 mm from the mouth of the tube. When fully exserted, the style has a length of up to 75 mm, which makes it the longest known for any South African aloe. The fruit capsules and winged seeds are also very large for such a small plant.

Flowering occurs in July and August.

When not in flower, *A. longistyla* may be confused with the superficially similar *A. humilis* or *A. brevifolia*. However, the flowers have several unusual features as described above and it is therefore unlikely to be confused with other species when in flower.

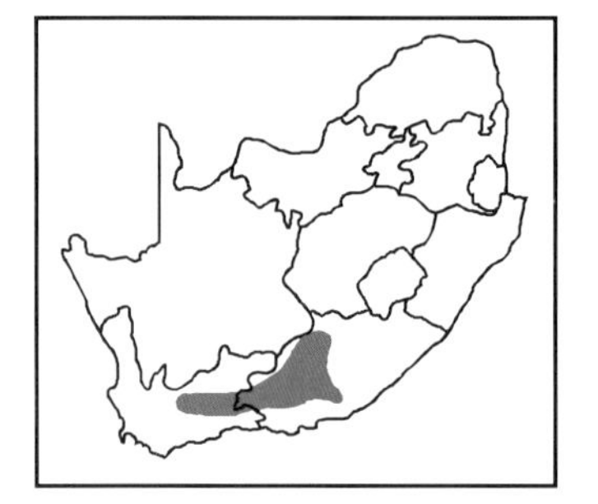

This dwarf aloe is widely distributed in semi-arid parts of the Little Karoo and south-eastern parts of the Great Karoo, from Calitzdorp in the west to near Grahamstown in the east, and extending northwards to Graaff-Reinet, Cradock and Middelburg. The plants are found scattered about, never occurring in dense groups as do most other aloes. They grow on flat stony or sandy areas or on gentle slopes, usually in the partial shade of small bushes. They are not easily spotted unless in flower.

Despite its wide distribution, *A. longistyla* is vulnerable. It is threatened by plant collectors and also by habitat degradation as a result of overgrazing.

Aloe longistyla does not grow easily in cultivation and plants should never be removed from their natural environment. In some parts of the distribution area, the species is known by the Afrikaans vernacular name Ramenas. The scientific name *longistyla* refers to the long style which protrudes from the mouth of the flower.

Ernst van Jaarsveld — NBI

Stewie Theron

Aloe variegata

Plants are stemless and up to 250 mm in height. The mottled leaves characteristically have a ridge or keel along the lower surface and are without spines or prickles. The green or brownish-green leaves are most commonly arranged in three distinct ranks and the margins have closely spaced small teeth along a white horny edge. The inflorescence is usually branched with relatively large hanging flowers varying in colour from dull pink to red, or rarely yellow.

The flowering period is from July to September.

This well-known aloe is very distinctive and unlikely to be confused with any other species. It can be distinguished from the related and exclusively Namibian *A. dinteri* and *A. sladeniana* on account of it being a more robust plant in all respects and having more densely flowered racemes. *A. variegata* is also more frequently encountered in cultivation than the other two Partridge-breasted aloes from Namibia.

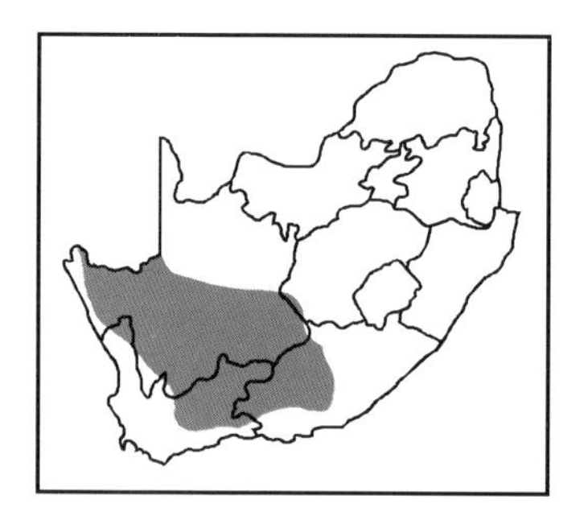

Aloe variegata is widely distributed in the dry regions of Southern Africa. It occurs in the Free State, the Karoo, Namaqualand and the southern parts of Namibia.

The species is not threatened.

Aloe variegata was one of the first aloes to be cultivated successfully in Europe. It is a striking plant with its finely spined, angular leaves which might cause one to confuse it with species of the closely related genus *Gasteria*. In some parts of the Karoo the plant is cultivated on graves. This practice may also be related to the Afrikaans common name, Kanniedood, suggesting eternal life, in addition to the more obvious explanation that the plants can withstand prolonged and extreme conditions of drought. The scientific name *variegata* refers to the variegated or mottled leaves. The species is fairly easy to cultivate, provided that it is given a well-drained soil mixture.

Forsyth van Nierop

Grass aloes

Group 10

This group of small, more or less stemless aloes is easy to recognise by the **long narrow leaves** that are **only slightly succulent**. The upper leaf surfaces are usually without spots but the **lower surfaces may have whitish spots**, especially towards the base. The inflorescences are invariably single, with a head-shaped or rounded raceme of white, pink, yellow, orange or red flowers. Most species have distinctive spindle-shaped roots; that is, they are swollen in the middle and taper towards the ends. Although the species are found in grassveld, the common name Grass aloe refers to the grasslike appearance of the plants and not to their preferred habitat. They often grow at high altitudes and flower mostly in summer rather than in winter, as do most other aloes. Fires do not kill grass aloes but will in fact often stimulate them to flower. When not in flower, the plants are generally difficult to distinguish from the grasses with which they grow.

SPECIES

A. albida
A. boylei
A. chortolirioides
A. cooperi
A. dominella
A. ecklonis
A. fouriei
A. hlangapies
A. inconspicua
A. integra
A. kniphofioides
A. kraussii
A. linearifolia
A. lmicracantha
A. minima
A. modesta
A. myriacantha
A. nubigena
A. parvifolia
A. saundersiae
A. soutpansbergensis
A. thompsoniae
A. verecunda
A. vossii

Ben-Erik van Wyk

Aloe verecunda

Aloe albida

Plants are stemless and occur singly or in small groups, with the leaves spirally arranged. The roots are swollen in the middle and taper to both ends (spindle-shaped). The leaves are narrowly linear and grasslike, up to 150 mm long and about 5 mm wide. They are without spots and have small white teeth along the margins. Inflorescences are simple and up to 180 mm high. The flowers are distinctly two-lipped, up to 18 mm long and dull white, with the tips of the petals and the veins greenish in colour.

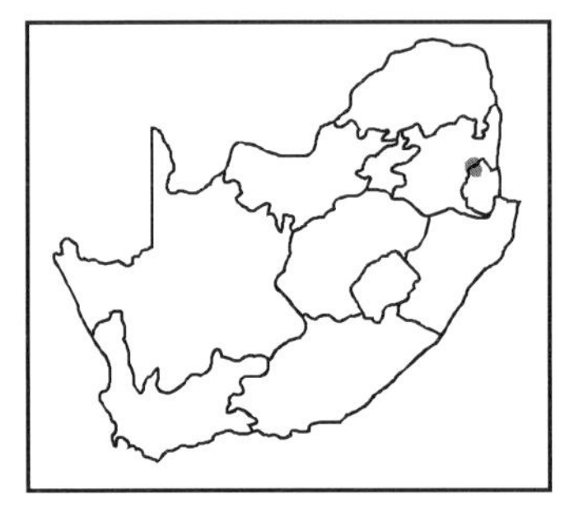

Flowering occurs in February and March.

Aloe albida is similar to *A. saundersiae* and *A. myriacantha* with its narrow, grasslike leaves. It differs, however, in the larger, two-lipped flowers that are white with green tips. The tips of the petals are curved upwards (the upper three more so than the lower three), resulting in the characteristic two-lipped appearance of the flowers.

This distinctive little aloe is found only in the Barberton area in Mpumalanga, where it occurs in the mist belt.

Aloe albida is not threatened but it has a limited distribution and should therefore be protected.

In cultivation, *A. albida* will grow well in rich moist soil. The scientific name *albida* means 'somewhat white' or 'whitish', in reference to the unusual flower colour.

Koos de Wet

Frits van Oudtshoorn

Aloe boylei

A. boylei is a robust, broad-leaved plant with short, thick stems of up to 200 mm long. The leaves are very large and up to 600 mm long and 100 mm wide at the base. The upper surface may have a few scattered, white spots, whereas the lower surface is copiously spotted near the base. The leaf margins have whitish borders with small, widely spaced teeth. The head-shaped inflorescences have relatively large flowers (about 40 mm long). The flowers are salmon-pink and tubular in shape, but taper off to the tips.

Flowering is from November to February.

Aloe boylei is superficially similar to *A. ecklonis* – both species have broad leaves which are uncharacteristic of grass aloes. However, *A. boylei* may be distinguished by its short stem, the more erect (not flatly spreading) leaves and the larger, tubular flowers.

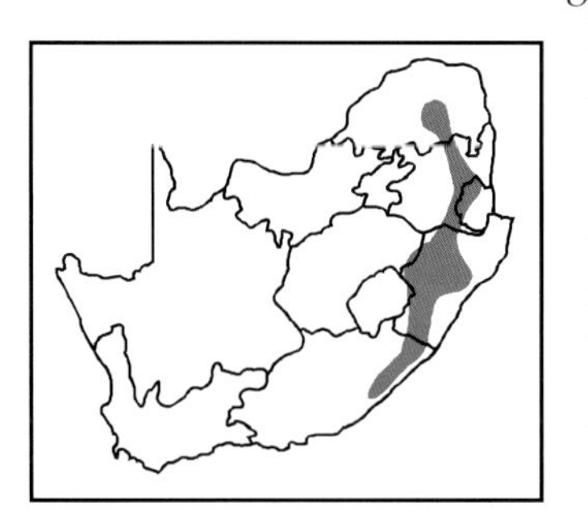

The species is widely distributed in the eastern parts of Southern Africa, from near East London in the Eastern Cape Province to Magoebaskloof in the north.

The species is not threatened.

As is the case with most species of grass aloes, the leaves die back in winter, leaving only a short thick stem from which new leaves emerge in spring. *Boylei* is derived from F. Boyle, the person who sent the first material to England, where the species was named. The Zulu names isiPutumane and isiPhukutwane have been recorded.

Neil Crouch

Aloe chortolirioides

Plants form dense tufts of up to fifty branches. The stems are somewhat woody and about 200 mm long. The leaves are erect, very narrow and grasslike, up to 250 mm long. There are often a few small spots on the basal parts of the lower surfaces and minute white teeth are present along the margins. Inflorescences are simple, single on each stem and up to 400 mm high, with rounded (head-shaped) racemes. The flowers are long, narrow, slightly bulging in the middle and up to 40 mm long. They are usually red but occasionally pink, orange or yellow.

Flowering may occur over a long period, from March to September. The peak flowering time is from June to September, and plants often flower in response to veld fires.

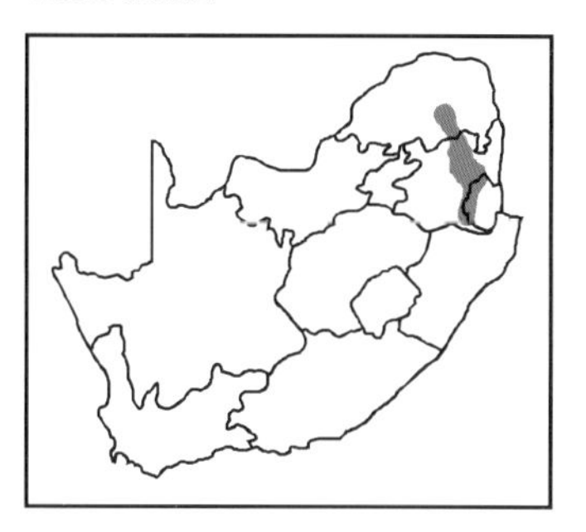

Aloe chortolirioides can be distinguished from other grass aloes by the branched and tufted growth form and the very narrow, grasslike leaves.

The wide distribution area includes the Northern Province, Mpumalanga and also Swaziland.

This grass aloe is not threatened.

In addition to the typical form, two varieties have been described: an orange-flowered form, and a robust form with broader leaves and slightly longer flowers, var. *woolliana*. The scientific name *chortolirioides* means 'resembling *Chortolirion*' (see *Aloe*-like plants, page 8).

Frits van Oudtshoorn

George Lubbers

Aloe cooperi

Plants occur singly or in small groups. The leaves are two-ranked or sometimes spirally arranged in old plants, and their bases are copiously white-spotted on the lower surfaces. The rather long, erect leaves have small white teeth along their margins. The inflorescences are invariably single from each rosette, and are usually taller than most other grass aloes, sometimes reaching more than 1,2 m. The racemes are cone-shaped and many-flowered. The long, tubular flowers are narrower towards the tips and are usually salmon-pink in colour but they may also be greenish, orange or reddish.

Like most other grass aloes, *A. cooperi* typically flowers in summer, from December to March. However, the recently described subsp. *pulchra* appears to flower slightly later in the season, from April to May.

Aloe cooperi can be distinguished from all other grass aloes by the ridged (keeled or V-shaped) lower surfaces of the leaves. Although this species may be confused with *A. micracantha*, it does not share the same geographical distribution. *Aloe micracantha* is essentially an Eastern Cape species. Furthermore, the inflorescences of *A. cooperi* are more cone-shaped than those of *A. micracantha*, which are head-shaped. *Aloe cooperi* subsp. *pulchra* may be distinguished from subsp. *cooperi* in that the margins in the upper three-quarters of the leaves are without teeth, while in subsp. *cooperi* the leaf margins are toothed throughout. In addition, the leaves of subsp. *pulchra* are more sharply keeled and thicker.

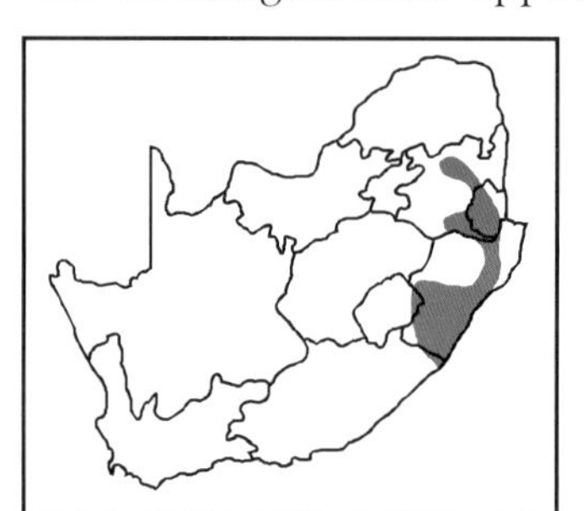

This grassland species occurs in dry rocky areas or in wet marshy habitats, mainly in KwaZulu-Natal and Mpumalanga.

Although the typical subspecies of *A. cooperi* is not threatened, subsp. *pulchra* has a localised distribution on the KwaZulu-Natal north coast and its conservation status is insufficiently known.

Aloe cooperi is not widely cultivated, but it is fairly easy to grow, even in cold areas. The species was first discovered by the famous explorer Burchell, but it was rediscovered by Thomas Cooper, after whom it was then named.

Leo Thamm

Geoff Nichols

Aloe dominella

The plants form branched tufts of up to fifty stems (usually ten to twenty). The leaves are narrow and grasslike, up to 350 mm long and 10 mm wide, with numerous spots on the basal part of the lower leaf surface. Minute white teeth are present along the leaf margins. Inflorescences are simple, up to 400 mm high, with head-shaped racemes. The flowers are very short, being only about 18 mm long and invariably yellow.

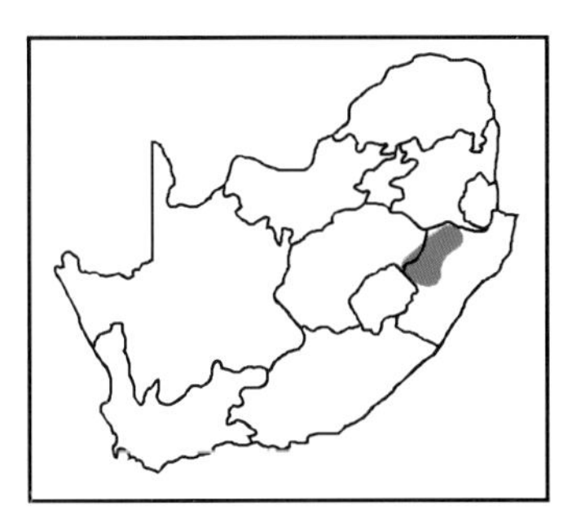

The flowering period is from July to October, depending on veld fires.

When not in flower *A. dominella* may be confused with *A. chortolirioides*, but can easily be distinguished by the short yellow flowers.

This species occurs in central and north-western KwaZulu-Natal.

Aloe dominella is not threatened.

The meaning of the name *dominella* is not clear, but it probably refers to the fact that the species is locally dominant in small areas.

Godfrey Symons

Geoff Nichols

Aloe ecklonis

Plants are broad-leaved and occur in medium to large clumps. Stems are usually absent or inconspicuous. The dull green leaves are 400 by 100 mm and arise from ground level. They have a few spots on the lower surface only and the margins have distinct white edges with small yet firm triangular teeth. The long, slender inflorescences are densely head-shaped with yellow, orange, salmon-pink or red flowers. The flowers are very short (about 20 mm long) and markedly swollen in the middle.

Flowering is from November to January.

Plants are fairly robust and the leaves are much broader and larger than in most other grass aloes. This species can be distinguished from *A. boylei*, another broad-leaved grass aloe, by its smaller and differently shaped flowers. (Other differences between the species are discussed under *A. boylei*.)

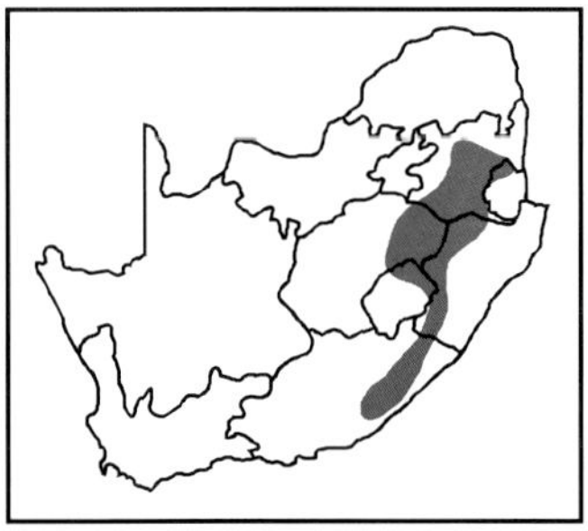

Aloe ecklonis has a wide distribution in the summer rainfall grasslands of the eastern escarpment.

The species is not threatened.

Although not widely cultivated as a garden ornamental, this species is very attractive because of the large number of inflorescences that are borne simultaneously. The scientific name commemorates the well-known plant collector C.F. Ecklon, who originally sent seeds of the species to Europe.

Cameron McMaster

Frits van Oudtshoorn

Piet Vorster

Aloe fouriei

Plants are stemmed, a characteristic which is not widespread amongst the grass aloes. The stems may reach a length of 150 mm. The leaves are also fairly long and slender, up to 300 mm by 25 mm. A few white spots occur on the upper leaf surfaces while the lower surfaces are more copiously spotted, particularly near the base. The inflorescences are unbranched and more or less 400 mm long, with dense heads of flowers. The flowers are broad at the bases and taper sharply to the narrow mouths. The tubular part of the flower is orange but the tips are bright green.

Flowering is mainly in summer, from November to February, but appears to vary in cultivation.

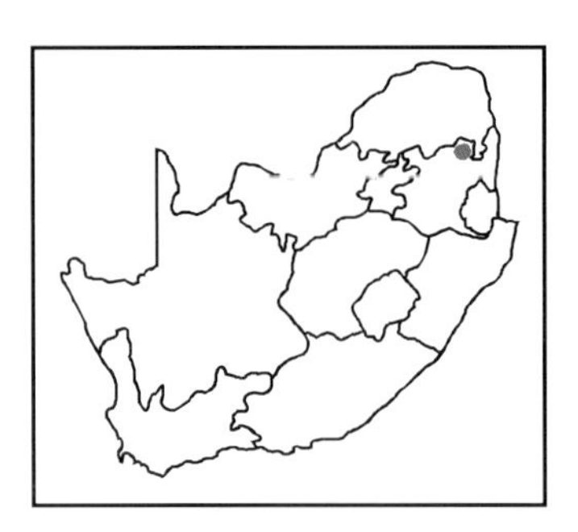

The flowers are very similar in colour to those of *A. nubigena* but can be distinguished by the different shape (narrow mouths). The more robust leaves of the species resemble those of *A. cooperi*, but the latter is a stemless plant.

The species occurs in Mpumalanga in the vicinity of Lydenburg and Pilgrim's Rest, where it is associated with dolomitic outcrops.

Aloe fouriei is only known from a few localities and the conservation status is regarded as insufficiently known.

This species was described fairly recently. It is little-known and not well-represented in cultivation. The species was named after a conservation officer, Steve Fourie, who first discovered it.

Koos de Wet

Koos de Wet

Aloe hlangapies

Plants are usually solitary but may also form small clumps. The dull green leaves are up to 500 mm long and 60 mm wide. They are often copiously white-spotted on the basal parts of the lower surfaces and have small triangular teeth along the margins. The inflorescences are invariably single and densely head-shaped, with yellow, orange or red flowers. The flowers are up to 30 mm long and somewhat swollen in the middle.

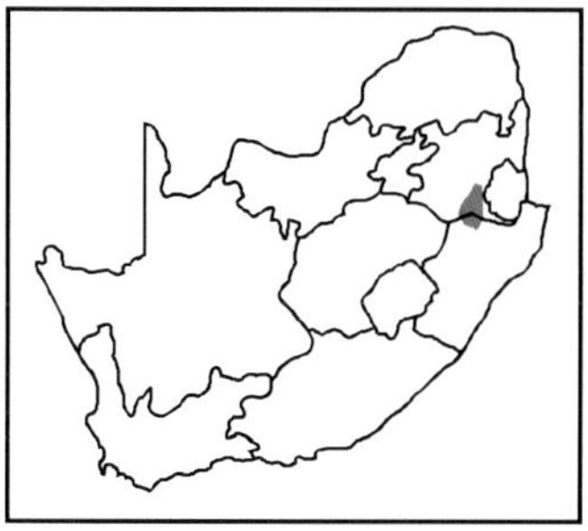

The species flowers in November and December.

Aloe hlangapies is very closely related to *A. ecklonis* and intermediates are often encountered. It can be distinguished by the fan-shaped leaves (not spirally arranged as in *A. ecklonis*), and by the longer, more tubular flowers.

The species occurs in grassland in the vicinity of Piet Retief and Vryheid.

Aloe hlangapies is not threatened.

It is with some doubt that *A. hlangapies* is here retained as distinct from *A. ecklonis*. *Hlangapies* is a Zulu place name that was directly used as a scientific name.

Frits van Oudtshoorn

Frits van Oudtshoorn

Aloe inconspicua

Plants are solitary, forming a bulb-like swelling below the ground. The leaves are long and narrow, with small marginal teeth. The inflorescence is a simple, cylindrical, densely flowered raceme. The erect, green and white flowers are stalkless and up to 15 mm long. The bracts in this species are almost as long as the flowers.

Aloe inconspicua flowers in November.

The species can be distinguished from *A. modesta* and *A. kniphofioides* (the other two grass aloes with bulb-like leaf bases) by its dense, cylindrical racemes of greenish flowers. The flowers are not scented as in *A. modesta*. In general appearance the flowers are like those of *A. albida*, but the latter lacks the dense cylindrical racemes and stalkless flowers.

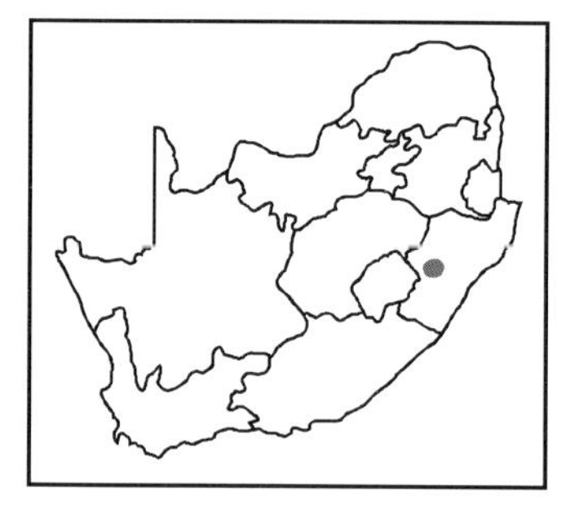

Aloe inconspicua grows on the transition between grassland and valley bushveld in the vicinity of Estcourt, KwaZulu-Natal.

The species is not threatened.

Aloe inconspicua was aptly named because it is inconspicuous and difficult to distinguish from the grass in which it grows. Consequently, it has remained undescribed until fairly recently.

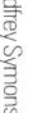

Godfrey Symons

Frits van Oudtshoorn

Aloe integra

The plants occur singly or in small groups of up to six stems. The fairly broad leaves are glossy green with narrow white margins. Marginal teeth are usually absent, but if present, they are small and inconspicuous. Inflorescences are up to 500 mm high and the racemes are usually rounded at the top. The cigar-shaped flowers are lemon yellow and are subtended by purple bracts.

Aloe integra flowers mainly in early summer, from October to December.

This is one of the more readily recognisable grass aloes. The short and relatively broad leaves usually end abruptly as a result of die-back caused by frost or veld fires. The tufted purple bracts at the tips of the racemes are also useful to distinguish the species from other yellow-flowered grass aloes.

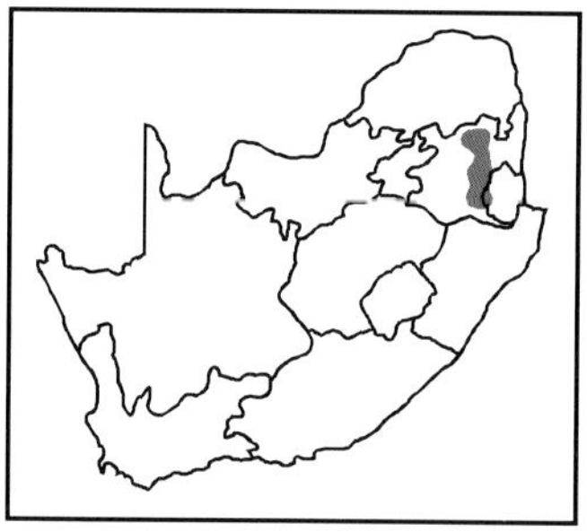

Aloe integra is distributed in a narrow zone from Swaziland northwards to the vicinity of Pilgrim's Rest.

Aloe integra is not threatened.

The species is very rarely found in gardens and does not flower readily in cultivation. *Integra* means 'without teeth', but this name is not very appropriate since the leaves generally do have small teeth along the margins.

GW Reynolds – NBI

NBI

Aloe kniphofioides

The plants are solitary and stemless, with the leaf bases widening to form an underground, bulb-like structure. The leaves are very narrow and grasslike, up to 300 mm long, without any spots, but often with minute white teeth along the margins. Inflorescences are simple, up to 500 mm high, with sparse oblong racemes of up to 150 mm long. The flowers are long, narrow and cigar-shaped. They are up to 50 mm long, with an attractive bright red colour. The tips of the petals are green, providing a striking contrast with the attractive bright red colour of the flowers.

The flowering time is in November and December.

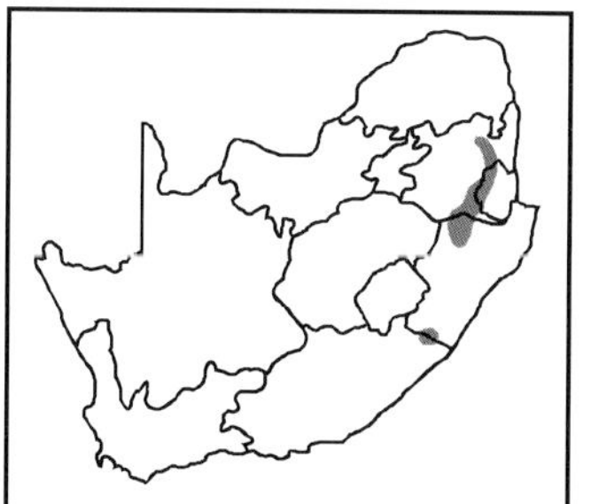

Aloe kniphofioides is easily distinguished from other South African aloes by the distinctive underground bulbs and also the long, narrow, bright red flowers. The only other South African aloes with bulbous bases are *A. modesta* and *A. inconspicua*, but these species have yellow or green and white flowers.

This interesting grass aloe is widely distributed in southern and northern KwaZulu-Natal, Mpumalanga and western Swaziland.

The species is not threatened.

The scientific name suggests a similarity with the genus *Kniphofia*, the well-known Red-hot pokers.

Bill van der Riet

Frits van Oudtshoorn

Duncan Butchart

Aloe kraussii

Plants are stemless and occur singly or in small groups. The dull green leaves are arranged in a fan-shaped rosette, but may become spirally arranged in old specimens. The leaves have minute teeth along the very narrow, white margins. Inflorescences of up to 400 mm high arise from each rosette. The racemes are broader than long, with relatively short, tubular lemon yellow flowers of which the mouths are slightly upturned.

Aloe kraussii flowers from November to February.

This species is very similar to *A. ecklonis*, but may be distinguished by the shorter, more tubular flowers of up to 18 mm long (those of *A. ecklonis* are broader and over 20 mm long). The leaves of *A. kraussii* are also narrower with smaller marginal teeth. In contrast to many other grass aloes, the leaves of *A. kraussii* are more or less devoid of spots on the basal parts of the lower surfaces.

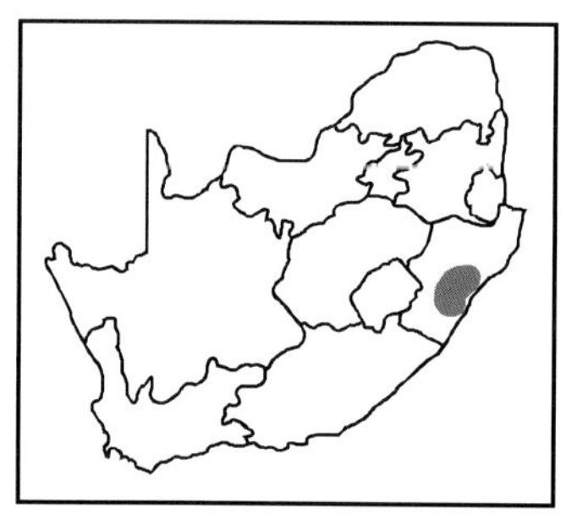

Aloe kraussii occurs on grassy hill slopes in central KwaZulu-Natal. Its occurrence in Swaziland needs to be confirmed.

The species is not threatened.

Aloe kraussii is closely related to *A. ecklonis* and may yet prove to be little more than a southern form of the latter. The species name commemorates the botanical collector C.F.F. Krauss.

Ben-Erik van Wyk

Ben-Erik van Wyk

Aloe linearifolia

This small aloe usually occurs as a solitary plant with long narrow leaves of up to 250 mm long and 10 mm wide. The margins of the leaves are without teeth or may have minute teeth near the base. As is the case with most other grass aloes, the lower surfaces are densely spotted at the base. The inflorescence is a single, head-shaped raceme with short, tubular flowers of up to 12 mm long. Flower colour varies from greenish-yellow to yellow.

Aloe linearifolia flowers in February and March.

The long, narrow leaves of this species distinguish it from other yellow-flowered grass aloes. It differs from the closely similar *A. kraussii* and *A. ecklonis* in the smaller size of the plants, narrower leaves and smaller flowers.

The species is found in the grasslands of southern and central KwaZulu-Natal.

Aloe linearifolia is not threatened.

Aloe linearifolia varies considerably in size and large, robust forms have been recorded in the southern part of its distribution.

Linearifolia means 'linear leaves', a botanical term used to describe leaves which are long and narrow – more than twelve times longer than wide.

Geoff Nichols

Aloe micracantha

Plants are generally single-stemmed and the grasslike leaves are spirally arranged. The leaves are fairly long and narrow, about 500 mm by 30 mm, and are copiously spotted on both surfaces. The inflorescences are unbranched, head-shaped racemes with relatively large flowers (more or less 40 mm long) that vary from red or orange to salmon-pink.

Aloe micracantha flowers in summer from November to February.

The slender leaves are generally U-shaped in cross-section (deeply channelled), and have numerous white spots on both surfaces. This is the only grass aloe that occurs in fynbos vegetation.

Aloe micracantha is restricted to the Eastern Cape Province and occurs from Uniondale in the west to Grahamstown in the east.

Due to urbanisation and agricultural development, the species is becoming increasingly rare in certain parts of its natural distribution range.

The name *micracantha* is derived from the small teeth on the margins, but this is by no means the species with the smallest marginal teeth. *Aloe micracantha* has very thick, long roots and mature specimens transplant with difficulty. It is therefore not wise to remove plants from their natural habitat. To cultivate the species successfully it is important to bear in mind its preference for poor, well-drained sandy soils.

GW Reynolds — NBI

Gideon Smith

Aloe minima

This species is found as single, stemless plants with spindle-shaped roots. The leaves are linear, up to 350 mm long and 4 to 6 mm wide, with numerous small spots on the basal parts of the lower surfaces. The margins have minute white teeth along the lower half.

The inflorescences are simple and up to 500 mm tall. The dull pinkish flowers are exceptionally small (10 to 12 mm long), with the tips of the petals slightly spreading, but not two-lipped.

The flowering time is from February to March.

Aloe minima may be confused with *A. saundersiae*, but it is a larger plant with taller inflorescences.

This species is widely distributed in KwaZulu-Natal, Mpumalanga and also southern Swaziland.

Aloe minima occurs over a wide area and is not threatened.

A robust form from Mpumalanga, with two-ranked leaves and inflorescences up to 600 mm high has been described as *A. minima* var. *blyderivierensis*. *Minima* means 'very little', referring to the small size of the plant.

GW Reynolds – NBI

Frits van Oudtshoorn

Aloe modesta

Plants are small and fairly inconspicuous. The leaves are long and thin and the basal portions form a bulb-like swelling below the ground. The inflorescences are characteristically densely packed with numerous, virtually stalkless flowers. The greenish-yellow and fragrant flowers are tubular and up to 13 mm long, and the mouths are somewhat curved upwards.

Aloe modesta flowers in January and February.

This species can easily be distinguished from other grass aloes by the swollen leaf bases, dense racemes and fragrant flowers.

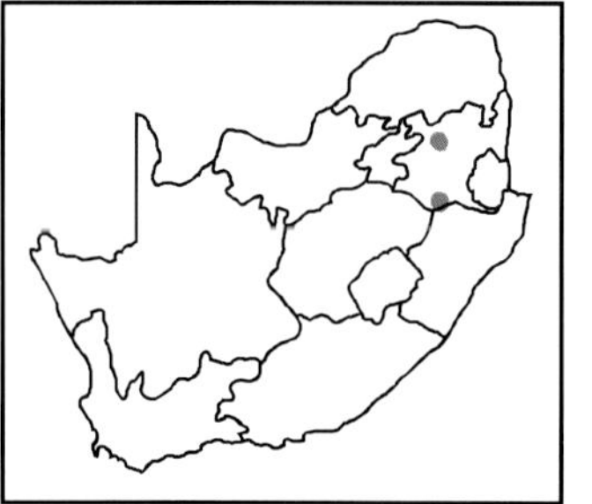

Aloe modesta is known from only two localities, one near Dullstroom in Mpumalanga and the other near Wakkerstroom on the border between Mpumalanga and KwaZulu Natal.

The conservation status of the species is insufficiently known.

Several species with scented flowers occur in Madagascar but this appears to be the only fragrant aloe of the African continent. The scientific name refers to the modest size of the plants.

Frits van Oudtshoorn

Ben-Erik van Wyk

Aloe myriacantha

The plants are stemless and invariably solitary, with spindle-shaped roots. The leaves are long and narrow, up to 300 mm long and 10 mm wide, with a few spots towards the bases and small white teeth along the margins. The inflorescences are simple and up to 300 mm high, with distinctly two-lipped flowers, the latter being up to 20 mm long and dull purplish-pink in colour.

Flowering occurs in March and April in Southern Africa, but somewhat later in central Africa. The flowers of this species are similar to those of *A. albida* as they are two-lipped in both species, but *A. myriacantha* is a larger plant and the flowers are usually dull purplish-pink, not white and green as in the latter.

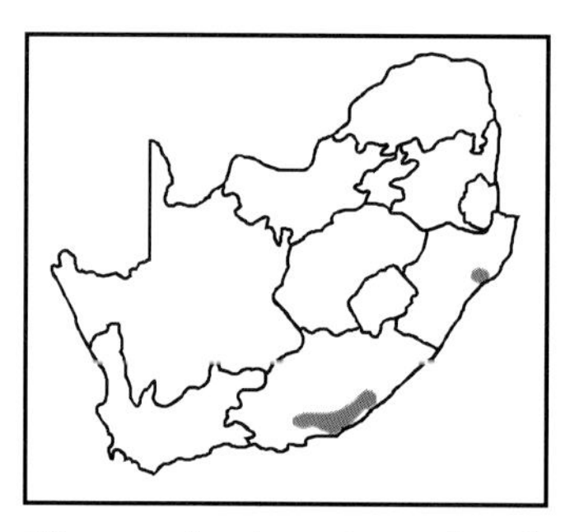

This species is widely distributed in the eastern parts of South Africa and occurs from near sea level up to more than 600 m.

Aloe myriacantha is not threatened.

The species is not restricted to South Africa, but also occurs in central Africa. The scientific name is derived from the Greek words *myri* meaning 'countless', and *acantha* meaning a 'spine' or 'prickle'. However, the leaves do not really have more thorns than do other grass aloes.

Ben-Erik van Wyk

Ben-Erik van Wyk

Aloe nubigena

The species is freely suckering and may form dense groups of up to twenty stems. The stems are up to 250 mm in length with long, narrow, often drooping leaves. The upper leaf surfaces are more or less flat and the lower surfaces are densely spotted near the bases, sometimes with a few scattered spots higher up. The margins have minute white teeth. The inflorescences are invariably unbranched with sparsely flowered, head-shaped racemes. The flowers are tubular, up to 250 mm long and vary from orange to red with green tips.

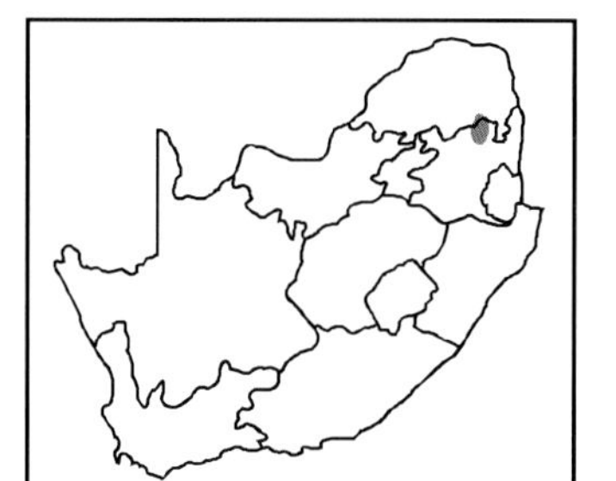

Aloe nubigena flowers from November to March.

The species is similar to *A. thompsoniae* and differences are given under the latter.

Aloe nubigena has a rather restricted distribution and occurs in rocky grassland along the escarpment in Mpumalanga.

The species is not threatened.

Aloe nubigena is easy to cultivate if given rich soil, and flowers freely if watered adequately. The species was named for its high-altitude distribution (*nubigena* means 'cloud borne').

Pitta Joffe NBI

Pitta Joffe – NBI

Aloe parviflora

This poorly known species has spindle-shaped roots and long narrow leaves which are covered with minute, soft spines on the lower surfaces and the margins. The inflorescences are simple and about 400 mm high. The flowers are exceptionally small (only 8 mm long), and pale pink, with a short straight tube.

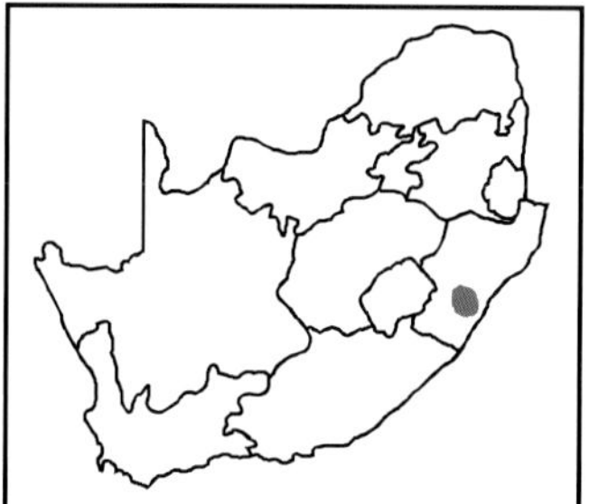

The flowering time is between January and March.

Aloe parviflora is closely similar to *A. minima* and may eventually prove to be merely a regional form of the latter. The minutely spiny leaves are distinct, but similar spiny tubercles are occasionally also found on the leaves of *A. minima*.

This species has been recorded from central KwaZulu-Natal, but the exact distribution is poorly known.

The conservation status of *A. parviflora* is indeterminate.

Aloe parviflora is superficially a very distinct species, but more information is required about the variation and natural distribution area. *Parviflora* is the Latin for 'small flowers'.

Ben-Erik van Wyk

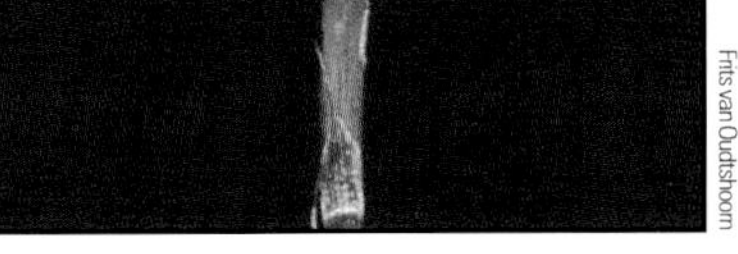

Frits van Oudtshoorn

Frits van Oudtshoorn

Aloe saundersiae

The plants occur singly or in small groups, and are stemless, with spindle-shaped roots. The leaves are narrow and linear, up to 100 mm long and about 3 mm wide, with or without a few spots near the bases and with minute triangular teeth along the margins. Inflorescences are simple and only approximately 180 mm high. The flowers are exceptionally small, up to 12 mm long, and dull pink in colour.

Flowering occurs in February and March.

The tips of the petals are slightly curved outwards, but the flowers are not two-lipped as in *A. albida* and *A. myriacantha*. It closely resembles *A. minima* in the size and shape of the flowers, but the latter is a much larger plant, with inflorescences of 300 to 500 mm high. Also, the leaves are erect, not spreading as in *A. saundersiae*.

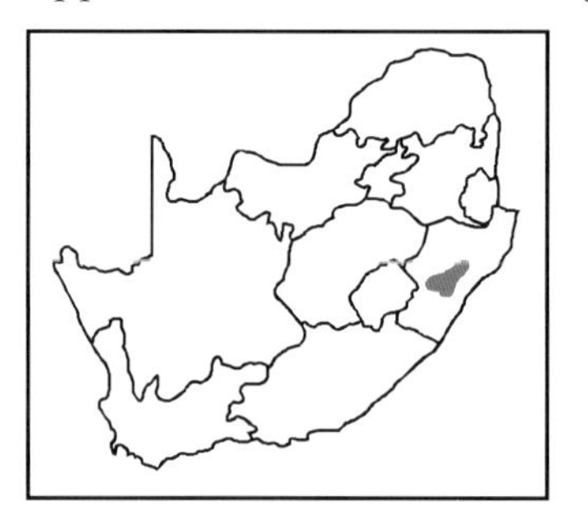

This species has a restricted distribution in the central part of KwaZulu-Natal.

Aloe saundersiae is not listed as threatened.

This species is perhaps the smallest of all aloes. It was named in honour of Lady Saunders, who first collected the species near the Nkandhla Forest in KwaZulu-Natal.

NBI

Pitta Joffe – NBI

Aloe soutpansbergensis

Plants usually occur in small clumps that consist of several rosettes. The long, narrow leaves are rather soft and droop to one side. The margins are curved inwards and have minute, translucent teeth. The simple inflorescences are up to 200 mm long, with a sparse raceme of tubular, orange flowers.

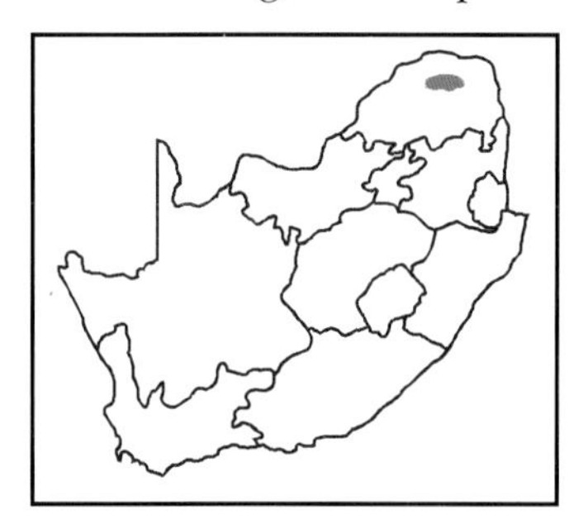

Aloe soutpansbergensis flowers in January.

This species can be distinguished by its large, orange flowers that may be up to 30 mm long. It appears to be midway between *A. verecunda* and *A. thompsoniae* in size and growth habit.

Aloe soutpansbergensis is a high-altitude, high-rainfall species that grows in the Northern Province on the Soutpansberg. What appears to be a robust form of the species has been collected on the Wolkberg.

The species is critically rare due to injudicious collection.

Aloe soutpansbergensis is surprisingly easy to cultivate and may in time form a large number of rosettes. The name was derived from its occurrence on the Soutpansberg.

Steve Fourie

Frits van Oudtshoorn

Aloe thompsoniae

Plants form small, dense groups through suckering. Rosettes are more or less stemless, with relatively broad succulent leaves. The leaves are spreading and recurved, about 200 mm long and 15 mm wide at the base. The upper leaf surface is hollow and gutter-like; the lower surface is rounded, with many small spots near the base. Some spots may also be present on the upper surface. The margins have minute white teeth, about 1 mm long. Inflorescences are simple, up to 200 mm high, with head-shaped to somewhat cone-shaped racemes. The red flowers are cylindrical, about 25 mm long, with the tips of the petals folded outwards.

The species flowers from December to early February.

Aloe thompsoniae is similar to *A. nubigena* but the latter has longer stems with the leaves arranged in rosettes (not in two ranks), the racemes are more densely flowered and the tips of the petals are recurved. *Aloe soutpansbergensis* is also similar, 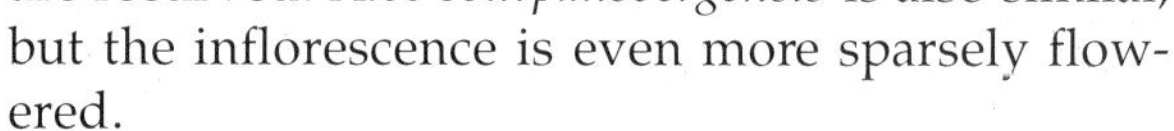
but the inflorescence is even more sparsely flowered.

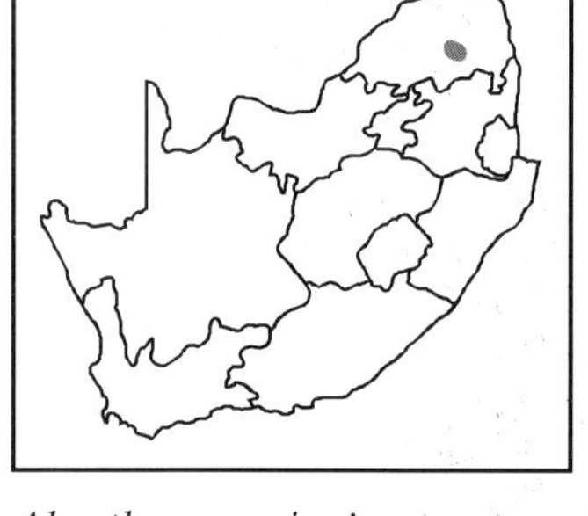

Aloe thompsoniae appears to be restricted to the Wolkberg in the Northern Province.

Even though this species is not threatened, the localised distribution area makes it vulnerable to overcollection.

Aloe thompsoniae is easy to cultivate, provided it is grown in rich soil in the shade and watered well. It was named after Mrs Thompson of Magoebaskloof, who first collected the species.

G.W.Reynolds – NBI

Pitta Joffe – NBI

Aloe verecunda

Plants are solitary or more commonly form dense groups from short stems that are branched at ground level. The leaves are usually fan-shaped but may occasionally be spirally twisted. The dull green leaves are hollow along the upper surface and have numerous tuberculate white spots along the lower surfaces. The inflorescences are invariably simple and the racemes are densely flowered, with rather long, tubular red flowers.

Aloe verecunda flowers from November to February.

Robust forms of this species may be confused with *A. cooperi*, but the latter species has keeled leaves and more cone-shaped racemes.

Aloe verecunda occurs widespread from the Witwatersrand in the south to Pietersburg in the north. It is usually found in rocky outcrops.

The species is not threatened.

Aloe verecunda is probably one of the best known of all the grass aloes because it is fairly common around Johannesburg and Pretoria. *Verecunda* means 'modest'.

Ben-Erik van Wyk

NBI

Aloe vossii

Plants are single or occur in small groups arising from short stems. The dark green leaves are long and narrow, about 30 mm wide at the base and taper gradually. Spots occur on the upper and especially the lower surfaces. The margins are curled inwards with firm, white triangular teeth along a thin white edge. The inflorescences are invariably single, up to 500 mm long with wide head-shaped racemes. The red flowers are tubular, but gradually become narrower towards the tips.

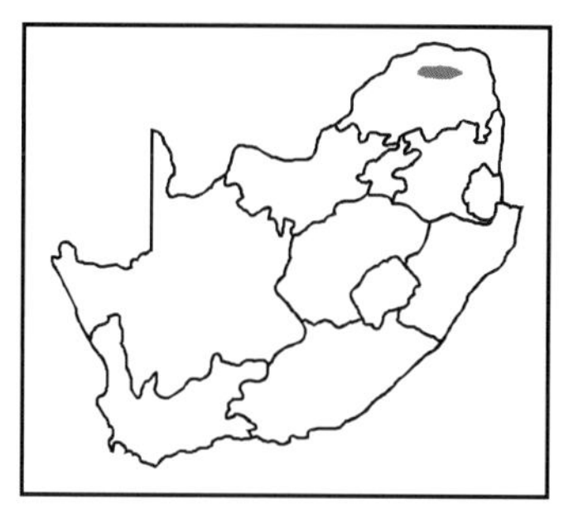

Flowering is from February to March.

Aloe vossii is similar to *A. verecunda*, but differs in having much longer leaves spotted on both sides. In addition, the flowers of *A. vossii* are not uniformly tubular, but are narrower towards the tips.

The species is restricted to the Soutpansberg.

Aloe vossii is critically rare.

Like most grass aloes this species has not been widely introduced into cultivation and remains little-known. It was named after the first collector, Harold Voss.

NBI

NBI

Further reading

Baker, J.G. 1896. Liliaceae, *Aloe*. In W.T. Thistelton-Dyer (ed.), **Flora Capensis**, Vol. 6,2. L. Reeve, Kent, pp. 302-329.

Beyleveld, G.P. 1973. **Aalwyne in die tuin.** Muller & Retief, Cape Town.

Bloomfield, F. 1985. **Miracle plants: *Aloe vera*.** Century Publishing, London.

Bornman, H. and D.S. Hardy. 1971. **Aloes of the South African veld.** Voortrekkerpers, Johannesburg.

Carter, S. 1994. Aloaceae. In R.M. Polhill (ed.), **Flora of Tropical East Africa.** A.A. Balkema, Rotterdam.

Court, G.D. 1981. **Succulent flora of southern Africa.** A.A. Balkema, Cape Town.

Groenewald, B.H. 1941. **Die aalwyne van Suid Afrika, Suidwes-Afrika, Portugees Oos-Afrika, Swaziland, Basoetoeland en 'n spesiale ondersoek van die klassifikasie, chromosome en areale van die *Aloe Maculatae*.** Nasionale Pers Beperk, Bloemfontein.

Gunn, M.D. and L.E. Codd. 1981. **Botanical exploration of southern Africa.** A.A. Balkema, Cape Town.

Jacobsen, H. 1974. **A handbook of succulent plants: descriptions, synonyms and cultural details for succulents other than Cactaceae. Vol. 1. *Abromeitiella* to *Euphorbia*.** 1st reprint. Jos. Adam, Brussels.

Jacobsen, H. 1977. **Lexicon of succulent plants: short descriptions, habitats and synonymy of succulent plants other than Cactaceae.** 2nd ed. Blandford Press, Poole, Dorset.

Jankowitz, W.J. 1975. **Aloes of South West Africa.** Division of Nature Conservation and Tourism, Administration of South West Africa, Windhoek. (Also available in Afrikaans and German.)

Jeppe, B. 1969. **South African aloes.** Purnell, Cape Town. (And various later updated editions.)

Jeppe, B. 1974. **Pride of South Africa: aloes.** Purnell, Cape Town.

Judd, E. 1972. **What aloe is that?** Purnell, Cape Town.

Reynolds, G.W. 1950. **The aloes of South Africa.** The Trustees of The aloes of South Africa Book Fund, Johannesburg. (And various later updated editions.)

Reynolds, G.W. 1966. **The aloes of tropical Africa and Madagascar.** The Trustees of The aloes Book Fund, Mbabane, Swaziland.

Smith, G.F. and Van Wyk, B-E. 1991. Generic relationships in the Alooideae (Asphodelaceae). **Taxon** 40: 557-581.

Van Jaarsveld, E.J. 1987. The succulent riches of South Africa and Namibia. **Aloe** 24: 45-92.

West, O. 1974. **A field guide to the aloes of Rhodesia.** Longman, Salisbury.

Index to scientific and common names

Bold = treated species. *Italic* = synonyms and other non-treated species. Normal type = common names.